Introduction to Elementary
MATHEMATICAL LOGIC

Introduction to Elementary
MATHEMATICAL LOGIC

Abram Aronovich Stolyar

Translated by Scripta technica, Inc.
Translation edited by Elliot Mendelson

DOVER PUBLICATIONS, INC.
NEW YORK

Copyright © 1970 by The Massachusetts Institute of Technology.
All rights reserved under Pan American and International Copyright Conventions.

Published in Canada by General Publishing Company, Ltd., 30 Lesmill Road, Don Mills, Toronto, Ontario.
Published in the United Kingdom by Constable and Company, Ltd., 10 Orange Street, London WC2H 7EG.

This Dover edition, first published in 1983, is an unabridged and unaltered republication of the work published by The MIT Press, Cambridge, Massachusetts, and London, England, in 1970. It was originally published in Russian as *Elementarnoye vvedeniye v matematicheskuyu logiku* by Prosveshcheniye Press, Moscow, in 1965. The present edition is published by special arrangement with The MIT Press, 28 Carleton Street, Cambridge, MA 02142.

Manufactured in the United States of America
Dover Publications, Inc., 180 Varick Street, New York, N.Y. 10014

Library of Congress Cataloging in Publication Data

Stoliar, A. A. (Abram Aronovich)
  Introduction to elementary mathematical logic.

  Translation of: Elementarnoe vvedenie v matematicheskuiu logiku.
  Bibliography: p.
  Includes index.
  1. Logic, Symbolic and mathematical. I. Title.
BC135.S7613 1983   511.3   83-5223
ISBN 0-486-64561-4

By virtue of its numerous applications in the most varied fields of science and technology (complicated problems in the foundations of mathematics, linguistic problems, certain problems of synthesis of automata, etc.), present-day mathematical logic is attracting the attention of a large number of persons in various specialties, including high school mathematics teachers.

The elements of mathematical logic and certain of its applications are included in the programs of high schools that place emphasis on mathematics. They can serve as interesting material for out-of-class work for students in the higher grades of any high school.

Recently, a number of monographs and collections of articles on mathematical logic have been published in the Soviet Union. However, there is almost no literature on mathematical logic available for high school mathematics pupils. The author of this book has set himself the task of providing an elementary exposition of the fundamentals of mathematical logic and certain of its applications for this broad class of readers.

The term "mathematical logic" encompasses at the present time an extremely broad and profound scientific study. In addition to classical logic of propositions and predicates, the fundamentals of which are expounded in this book, it includes a number of other logical systems and theories (constructive logic, systems of modal logic, many-valued logics, the theory of algorithms, etc.). For this reason, this book cannot be regarded as an introduction to mathematical logic as a whole. However, a familiarity with the material expounded in it will make it easier for the reader who is seriously interested in studying the subject to read the literature recommended for this purpose at the end of the book.*

The book includes exercises for independent work on the part of the reader.

The author expresses his deep appreciation to Yu. A. Gastev, whose suggestions have proved of great help in the preparation of the manuscript.

---

* The references to this bibliography are indicated by brackets enclosing the number of the reference in question.

Author's preface	v
INTRODUCTION	1

1
PROPOSITIONAL LOGIC 29

1. Objects and operations. 30
2. Formulas. Equivalent formulas. Tautologies. 46
3. Examples of the application of the laws of the logic of propositions in derivations. 62
4. Normal forms of functions. Minimal forms. 72
5. Application of the algebra of propositions to the synthesis and analysis of discrete-action networks. 82

2
THE PROPOSITIONAL CALCULUS 95

1. The axiomatic method. The construction of formalized languages. 96
2. Construction of a propositional calculus (alphabet, formulas, derived formulas). 101
3. Consistency, independence, and completeness of a system of axioms in the propositional calculus. 115

3
PREDICATE LOGIC 127

1. Sets. Operations on sets. 128
2. The inadequacy of propositional logic. Predicates. 136
3. Operations on predicates. Quantifiers. 143
4. Formulas of predicate logic. Equivalent formulas. Universally valid formulas. 156
5. Traditional logic (the logic of one-place predicates). 169
6. Predicate logic with equality. Axiomatic construction of mathematical theories in the language of predicate logic with equality. 178

APPENDIXES 189

I. A proof of the duality principle for propositional logic. 190

II. A proof of the deduction theorem for the propositional calculus. 191

III. A proof of the completeness theorem for the propositional calculus. 193

BIBLIOGRAPHY. 199

INDEX OF SPECIAL SYMBOLS. 201

INDEX. 203

Introduction to Elementary
MATHEMATICAL LOGIC

INTRODUCTION

1. Mathematical logic is essentially present-day formal logic. The ancient Greek philosopher Aristotle (384–322 B.C.) is considered the founder of formal logic. He was the first to develop the theory of deduction, that is, the theory of logical derivation, and he also discovered the formal nature of logical derivation, namely, that in our reasoning, certain propositions are derived from other propositions on the basis of a definite connection between their form or structure, independently of their specific content.

Our reasoning regarding two things of completely different specific content and applied in different branches of science or everyday life can have exactly the same structure, exactly the same form. For example, consider our reasoning with regard to the following three statements:

a. A square is a rhombus and a rhombus is a parallelogram; consequently, a square is a parallelogram.

b. A natural number is an integer and an integer is a rational number; consequently, a natural number is a rational number.

c. An oak is a tree and a tree is a plant; consequently, an oak is a plant.

As one can easily see, our reasoning had exactly the same form (structure) in all three cases. In each case, from two propositions (premises), we derive a third (the conclusion), the one appearing after the word "consequently." Also, in all three cases the validity of our drawing the conclusion from the premises is determined not by the specific nature of the subject but by the form of the premises and the conclusion, which was the same in all three cases.

Formal logic also studies the forms of human reasoning without paying attention to their specific subject matter. It seeks the answer to the question, How do we reason?

Aristotle's logic has been supplemented, modified, and improved in the course of many centuries by various philosophers and philosophical schools. However, it was not until the nineteenth century, when logicians began to apply mathematical methods, that really significant progress was made in this science. Formal logic became "truly formal" only when they began to study meaningful logical thinking by representing it in formal systems, logical calculi.

2. The idea of the possibility and desirability of mathematical logic was proposed as far back as the seventeenth century by the German mathematician and logician G. W. Leibniz (1646–1716). His attempts were the first in the history of science to represent logic in the form of an algebraic calculus.

However, not until the middle of the nineteenth century did mathematical logic begin to develop as a special branch of science. It arose as the result of application of mathematical methods to problems of formal logic, and its early growth was primarily a consequence of the efforts of the English mathematician George Boole (1815–1864) (*Mathematical Analysis of Logic*, 1847; *Investigation of the Laws of Thought*, 1854). Boole applied to logic the methods of the algebra of his day; that is, the language of symbols and formulas, and the setting up and solution of equations. In the works of Boole and another English mathematician, A. de Morgan (1806–1878), mathematical logic was developed as a special algebra—the algebra of logic.

The algebra of logic, which was interpreted primarily as the algebra of classes (sets) and then as the algebra of propositions, was further developed and defined in the works of the English scholar W. S. Jevons (1835–1882), the German mathematician E. Schröder (1853–1901), and the Russian scholar P. S. Poretskiy (1846–1907).*

Letter symbols in logic were even used by Aristotle, though to a very limited extent since, even in mathematics, a symbolic language, which could have been borrowed, had not been developed. Formal logic became truly symbolic only when the symbolic language developed in mathematics was applied to it. This justifies the

---

* For more detailed information on the history of the beginnings and development of mathematical logic, the reader is referred to the following books: William C. Kneale and Martha Kneale, *Development of Logic* (New York: Oxford University Press, 1962); and I. M. Bochenski, *History of Formal Logic* (Notre Dame, Ind.: University of Notre Dame Press, 1961).

term "symbolic logic" which is often used as a synonym for the term "mathematical logic."

A precise and clear (unambiguous) symbolism is above all necessary for clarification of the logical structure of our reasoning. Our everyday language is poorly adapted to this. To see this, consider the following two sets of statements written in everyday language:

a. I have met P. R. Popovich;
   P. R. Popovich is a cosmonaut.
   Consequently, I have met a cosmonaut.

b. I have met someone;
   someone has invented the radio.
   Consequently, I have met the inventor of the radio.

These two sets of statements have the same form from a standpoint of language, though the logical deduction is valid only in Statement a. The linguistic form of these sets of statements is inadequate for their logical structure.

It has become necessary to perfect the language of logic in such a way as to make it impossible to express lines of reasoning of different structure in the same linguistic form, in other words, to make the linguistic form determine unambiguously the logical structure of the reasoning. The symbolic language of mathematics, with a broad application of different types of variables, proved suitable for this purpose. Thus, the application of the precise mathematical language of symbols and formulas to logic was not happenstance but a necessity reflecting the requirements of logical language and, just as in mathematics itself, this application led to its remarkable development.

3. The development of the first non-Euclidean geometrical system by the Russian mathematician N. I. Lobachevski (1793–1856) and the Hungarian mathematician J. Bolyai (1802–1860) served as a stimulus to extensive investigations into the foundations of mathematics. The subject of these investigations was

primarily the axiomatic method of construction of mathematical theories.

There arose a new branch of science dealing with the study of the construction of mathematical theories. Hilbert called this study **metamathematics** or the **theory of proof**. The subjects of the study of metamathematics are not the sets, numbers, functions, etc., which are studied in mathematics, but the methods of operating with them in mathematics. The subject matter of metamathematics is mathematical theory itself. (In general, a theory describing the construction and properties of some other theory is called the **metatheory** corresponding to that other theory; the theory studied by the methods of the metatheory is called in this case the **subject**, or **object**, theory.)

As we know, the axiomatic method sprang up in geometry under the influence of the Euclidean "postulates," which for almost two millennia were considered unsurpassed in rigor as foundations of geometry, although from a present-day point of view this is, of course, far from the case. The system of initial propositions (postulates and axioms) proposed by Euclid proved insufficient as a basis for a logical development of all geometry. Furthermore, the logical means of drawing conclusions that are used in the construction of geometric theory were not made sufficiently precise. For example, when one said that in the axiomatic construction of a theory the propositions of this theory were derived by a purely logical procedure from certain initial propositions (axioms), it was not made precise what the phrase "by a purely logical procedure" meant, although this seemed intuitively clear.

The appearance of Lobachevski's non-Euclidean geometry (1826), the propositions of which, in contrast with the propositions of Euclidean geometry, are incompatible with the "usual" spatial representations, stirred up a number of complicated problems, notably the problem of consistency of the axioms.*

---

* For more detail on these problems, see [5], Part I, and [8], Introduction. See also Kneale and Kneale, *Development of Logic* and R. Blanché, *Axiomatics*, translated by G. B. Keene (New York: Dover Publications, Inc., 1962).

Investigation of these problems was impossible with the aid of the imperfect, insufficiently formalized apparatus of traditional logic. This circumstance served as a stimulus for the development of mathematical logic and influenced its application to the problems of the foundations of mathematics. This application in turn necessitated further development of mathematical logic, the introduction into it of a number of new ideas and methods, and the development of new logical calculi.

A new stage in the history of mathematical logic is associated first of all with the name of the German mathematician and logician G. Frege (1848–1925). Frege attempted to construct a perfect logical language that is suitable for mathematics. In his chief work *Grundgesetze der Arithmetik* (two volumes, 1893 and 1903), he constructed the first logical-mathematical system in the history of science which included a considerable amount of arithmetic.

Many outstanding mathematicians and logicians of the late nineteenth and twentieth centuries participated in the development of mathematical logic and its application to the theory of mathematical proof. These include the Italian mathematician G. Peano, the English scholars B. Russell, A. Whitehead, and A. M. Turing, the Polish logicians J. Lukasiewicz and A. Tarski, the German mathematicians D. Hilbert, W. Ackermann, and G. Gentzen, the Austrian mathematician K. Gödel, the Swiss mathematician P. Bernays, the French mathematician J. Herbrand, the Norwegian mathematician T. Skolem, the Dutch mathematicians L. E. J. Brouwer and A. Heyting, the American mathematicians and logicians S. C. Kleene, E. L. Post, J. B. Rosser, A. Church, and W. V. O. Quine, the Soviet mathematicians A. N. Kolmogorov, P. S. Novikov, A. A. Markov, and N. A. Shanin, and their numerous pupils.

Mathematical logic has made possible the perfection of the axiomatic method and it itself has been perfected with the aid of this method.

4. As can be seen from Sections 2 and 3, the phrase "mathematical logic" can be interpreted in two ways.

On the one hand, this branch of science is constructed as a

mathematical theory and mathematical methods are used in it, so that, in this sense, it can be thought of as "mathematical logic"; on the other hand, in developing the precise logical language of mathematics, it serves as the "logic of mathematics." A careful analysis of the relationship between logic and the subject matter of mathematics involves some profound philosophical problems which cannot be the subject of this book.*

5. We already stated (Subsection 2, p. 4) that one of the characteristic features of mathematical logic is the use of the mathematical language of symbols and formulas.

In mathematical language, just as in everyday language, we use the *names* of objects, that is, conventional linguistic expressions that are used to denote these objects,† making the distinction in so doing between the name of the object and the object itself that is denoted by that name.

Thus, we distinguish the number five as the common characteristic ("invariant") of an equivalance class of sets, for example, the set of fingers on the human hand, from the word "five" which denotes this number in English, or from the Russian word "pyat'," or from the symbols 5, 101 (the representation of this number in the binary system), V, etc., that are used to denote it in various numeration systems.

A language is well adapted to a precise description of a certain class of objects if in that language the following two conditions are satisfied: (1) for each object, there is a name for the properties of the object and the relationships between the objects of that class; (2) different objects, their properties, and their relationships have different names. If the first of these conditions is not satisfied, the language is poor and insufficient for describing the given class of objects; on the other hand, if the second condition is not satisfied, the language is ambiguous. For various reasons, influenced by history, the natural languages possess such ambiguity. The presence of homonyms, that is, words identical in form that

---

* See P. Benacerraf and H. Putnam, *Philosophy of Mathematics, Selected Readings*, Prentice-Hall, 1964.
† For more details regarding names, see [1] (Introduction, Section 1).

serve as names for distinct objects (an example would be the word *rest* in the sense of "repose" and in the sense of "remainder") is a violation of the second condition. Mathematical language, which grew out of our everyday language, is the result of the refinement of that language, in particular, the elimination of ambiguity.

We do not as a rule make the requirement on a language that distinct names denote distinct objects; that is, we allow synonyms. Generally speaking, this applies also to mathematical language. For example, we may consider $1 + 2$ and 3 as different names for the same number.

In elementary algebra, we use letters for the most part to denote numbers. In logic, we use letters to denote logical objects, for example, statements. By a **statement**, we mean that which is usually understood by this term in the grammar of any natural language, specifically, a linguistic expression or a combination of words that have independent meaning. (When we apply the term "statement" in what follows, we shall always mean statement in this sense.)

In the course of reasoning (not only in mathematics), we start with certain statements and from them formulate others, transforming them with the aid of the word *not* or joining them with the aid of connectives such as *and*, *or*, (*if. . ., then*), and *if and only if*, which indicate certain logical relationships between the statements.

To clarify the structure of the compound statements constructed in this way, it is convenient to denote the initial statements by letters without regard to their specific content. For example, consider the following statement:

If a quadrilateral is a parallelogram *and* its diagonals bisect the angles *or* are perpendicular to each other, *then* this quadrilateral is a rhombus.

This statement (called in grammar a compound statement) is constructed with the aid of the logical connection expressed by the words "if. . ., then" from the two statements

A quadrilateral is a parallelogram *and* its diagonals bisect its
angles *or* are perpendicular to each other (1)
and
This quadrilateral is a rhombus. (2)

Statement 2 is simple; that is, it does not admit further decomposition into other statements. Statement 1, which is compound, is constructed with the aid of the connective *and* from the two statements

A quadrilateral is a parallelogram (3)
and
Its diagonals bisect its angles *or* are perpendicular to each
other. (4)

Statement 3 is simple. Statement 4 is compound: It is constructed with the aid of the connective *or* from the two statements

Its diagonals bisect its angles (5)
and
(The diagonals) are perpendicular to each other. (6)

Statements 5 and 6 do not admit further decomposition.

Let us denote
Statement 3 by the letter A,
Statement 5 by the letter B,
Statement 6 by the letter C,
Statement 2 by the letter D.

In other words, we are assigning names to these statements.
Then, our compound statement can be written

If A and B or C, then D. (7)

However, this way of writing can be understood in two different ways if we do not adopt some additional conventions—it is not clear just what statements are *directly* connected by the connectives *and* and *or*. This defect can be eliminated by introducing parentheses. These parentheses can be put in different places, and where we put them determines different structures of the compound statement:

If A and (B or C), then D, (8)

If (A and B) or C, then D. (9)

By taking into consideration the specific content of the statements A, B, C, and D and the meaning of the logical relationships expressed by the words *and*, *or*, and *if*..., *then* (which will be the subject of our study in what follows), one can easily see that Statement 8 is true and Statement 9 is false. (It might be noted that Statement 7 without the parentheses is naturally read like 9; that is, it is natural to assume, as in ordinary arithmetic and algebra, that the parentheses are used only to indicate deviations from the "natural" way of reading a statement.)

One of the important features of mathematical language is that it allows the application of *variables* of different kinds, and so this language is suited for the expression of abstract *forms* that can be filled in by various specific items. This idea is essentially well known to all of us. We encounter it frequently in filling out various standard blank forms.

For example, suppose that a secretary is instructed to take a sample of a library card given to one student and to type up a "form" for all members of the class, following the outline of that one card. To obtain such a form, the secretary omits certain words and parts of words relating to the particular pupil and leaves blank spaces at those points. The result may look like this:

LIBRARY CARD

Given to _____, certifying that $\begin{Bmatrix} he \\ she \end{Bmatrix}$ is a pupil in the ninth grade of High School No. 444 in Moscow. To be presented in the district library.

_____

Principal

_____, 19___.

Of course, it is only when these blank spaces are filled in properly that this actually is a library card. But what do we mean when we say, "filled in properly"?

Each of the blank spaces has its own purpose, which can easily be guessed from the context. The first blank space is filled in with the name of the pupil. In the braces, the word "he" is struck out if the pupil is a girl and the word "she" is struck out if the pupil is a boy. Finally, at the bottom, the principal fills in his name and the rest of the date.

It is only when these conditions are strictly observed that the form can be considered filled in properly. If it is filled in improperly, the card is invalid or false. (For example, this last is the case when the first blank space is filled in with the name of a person who is not in the ninth grade of the school mentioned.)

If we wish to make up a library card form not merely for pupils in the ninth grade but for pupils of any grade in High School No. 444, we would have to omit the word "ninth" in the sample form. In the new form, there would be one more blank space, to be filled in by "first," "second,"..., "tenth," depending on the grade of the particular pupil in question.

If we omit the number of the school, we have a form with yet another blank space which can be used for a library card for a much broader class of pupils, since one can then fill in the number of any school in Moscow (an arbitrary element of the set of these school numbers).

In mathematics, one can use blank spaces only in the very simplest cases. For example, we can write

$$3 + \underline{} = 10,$$

but, if we do this, we need to indicate what may be put in the blank space since here, in contrast with the library card form, one cannot guess the set of numbers whose names we can put in this space, and this is quite important. (Henceforth, we shall, for brevity, say that we may fill in the numbers of such and such a set, meaning actually the names of these numbers.)

For example, if we are permitted to put in this blank space only numbers belonging to the set of negative integers, then there is not a single one that will render this form a true equation. On the other hand, if we are permitted to put natural numbers in this space, then such a number does exist.

However, for the expression of general mathematical laws, a language "with blank spaces" is unsatisfactory.

For example, we do not know how to "fill in" the "form"

$$\underline{} + \underline{} = \underline{} + \underline{},$$

which expresses, let us assume, the commutative law of addition for the set of integers. We do not know in what pairs of blank spaces we need to put a single number and into what pairs we may put different numbers.

In mathematical language, these difficulties are overcome by the use of *letters*. All the blank spaces into which we must put the same number are denoted by a single letter. These letters, known as **variables,** play the role of blank spaces. A variable is not the name of any *particular* element of any set. Instead, in mathematical usage, it is placed in that position which we are permitted to fill in with an *arbitrary* element of some *specified* set.

If we analyze this question carefully, it becomes clear that, in everyday speech, the role of "blank spaces" or "variables" is actually played by words denoting a *generic* concept (rather than an individual object). Everything that is said below regarding variables in the symbolic language of mathematics and mathematical logic holds for sentences in everyday speech as, for example,

The man plays chess well.

This sentence becomes a *proposition* (regarding which it is meaningful to say that it is true or false) only when we substitute for the generic variable "man" the name of some *particular* man. An analogous role is played by the indefinite pronouns (or even personal pronouns such as "he" in the sentence "he is a braggart" if we do not know the preceding context), indefinite articles, etc. In short, a variable is an "indefinite name" referring to an *arbitrary* object belonging to some set.

For example, if we write

$$x + y = y + x \quad \text{for all } x, y \in C,$$

this expresses the commutative law of addition for the set of integers with the aid of the variables denoted by the letters x and y. In the phrase "for all x, $y \in C$," where C is the set of integers and the symbol $\in$ is the symbol for membership, x and y are variables for which we may substitute the names of arbitrary integers.

Letters are not the only symbols that can be used as variables. For example, commutativity of any operation (it is immaterial what operation) can be indicated as follows:

$$x * y = y * x \quad \text{for all } x, y \in G.$$

Here, not only are x and y variables, but the symbol $*$ is also a variable, in place of which we can substitute the symbol for the specific operation in question ($+$, $\times$, etc.). Here, even G can be considered as a variable, in place of which we can substitute the name of some specific set.

Thus, by a variable we mean a symbol in place of which we can substitute the names of elements of some set.

The objects whose names we are permitted to substitute for a variable are called its **values** and the set of these objects is called the **range of values** of that variable (see p. 12).

In elementary algebra, we use only variables of a single type, namely, variables whose values are numbers. Such variables are called **numerical variables**. With the aid of these variables, the names of specific numbers, and the symbols for operations, forms are set up that are converted into numbers when we replace the variables with their values. For example, $(x + y) \cdot z + 2$ is a form for a number. (In elementary algebra, this form is usually called an **algebraic expression**.)

Suppose, for example, that the range of values of the variables x, y, and z is the set of natural numbers. Let us replace these variables with certain of their values, let us say, with 1, 2, and 3, respectively. We obtain $(1 + 2) \cdot 3 + 2$, that is, 11.

If the form contains one of the symbols $=$, $<$, or $>$ (meaning "is equal to," "is less than," and "is greater than," respectively), it no longer is a form for a number.

For example, if in the form $x < 3$, we replace the variable x with its value 2, we obtain not a number but the true statement $2 < 3$. If, instead, we replace x with 3, we obtain the false statement $3 < 3$. If, in the form $x^2 + 2 = 3x$, we replace x with 1, we obtain not a number but the true statement $1 + 2 = 3 \cdot 1$. If instead we replace x with 4, we obtain the false statement $16 + 2 = 12$. For such forms ($x < 3$, $x^2 + 2 = 3x$, etc.) containing variables and a symbol for some relation ($<$, $=$, $>$), we shall retain the term "statement."

We shall call a statement about which it is meaningful to say that its content is true or false a **proposition**. A statement containing a variable is not a proposition. For example, the statement "$x < 3$" is obviously not a proposition since it is not meaningful to say that it is true or that it is false. However, if we replace the variable x by any number (its "value"), we do obtain a proposition—a true or a false propostion depending on which number we substitute for x.

The statement "$x^2 + 2 = 3x$" is also not a proposition. We cannot say that it is true or that it is false since x is a variable. On the other hand, if we replace x by any of its values (in a given range), we obtain a proposition. This proposition is true or false depending on which value of the variable we substitute for x, as is clear from the following table:

x	$x^2 + 2 = 3x$	Type of proposition
0	2 = 0	F (False proposition)
1	3 = 3	T (True proposition)
2	6 = 6	T
3	11 = 9	F
4	18 = 12	F

The numbers 1 and 2 convert the statement "$x^2 + 2 = 3x$" into a true proposition; any other number substituted for x converts it into a false proposition.

Up to now, we have spoken only of numerical variables, with which we are familiar from elementary algebra. In mathematical logic, we also use other types of variables whose values are not numbers but logical objects—propositions are an example of this. The necessity of using such variables arises in logic, for example, in deciding whether one compound proposition follows from another.

By a **compound proposition**, we mean a proposition that can be decomposed into other propositions. If no part of a proposition is itself a proposition (or, at any rate, we do not consider it as such), we call it an **elementary proposition**.

Suppose that we are given a triangle ABC and a point D on the side AC. From the propositions "$AB = BC$," "$BD \perp AC$," and "$AD = DC$" we can construct various compound propositions. Let us look at some of these:

If $AB = BC$ and $BD \perp AC$, then $AD = DC$. (1)

If $AB = BC$ or $BD \perp AC$, then $AD = DC$. (2)

If $AB = BC$ and $AD \neq DC$, then BD is not perpendicular to AC. (3)

If $AD \neq DC$, then $AB \neq BC$ or BD is not perpendicular to AC. (4)

One can easily see that Proposition 1 is true and Proposition 2 is false. Yet these propositions differ only in the fact that the logical connective *and* in Proposition 1 is replaced in 2 by the connective *or*. As one can see, the truth of a compound proposition is determined not only by the truth of the propositions comprising it but also by the set and order of logical connectives with the aid of which this compound proposition is formed from the elementary propositions. (The proposition "if $AB = BC$, then $BD \perp AC$ and $AD = DC$" differs from Proposition 1 only in the order of the logical connectives and yet it is false.)

The set and order of the logical connectives (or operations) which form a compound proposition from elementary propositions constitute the logical structure of the compound proposition.

Propositions 1 and 2, although composed of the same elementary propositions, have different logical structures, and it is just this difference in structure that causes the difference in the truth values of these propositions, that is, the fact that 1 is true and 2 is false.

The structures of the compound propositions 3 and 4 also differ from the structure of Proposition 1, but these, like Proposition 1, are true; in fact, their truth *follows* from the truth of Proposition 1 independently of the content of the elementary propositions contained in them.

Let us look at the following three compound propositions:

If $a > b$ and $b > 0$, then $a > 0$, $\qquad$ (1′)

If $a > b$ and $a \not> 0$, then $b \not> 0$, $\qquad$ (3′)

If $a \not> 0$, then $a \not> b$ or $b \not> 0$, $\qquad$ (4′)

where a and b are numbers belonging to some set.

What is there in common between propositions 1 and 1′, between 3 and 3′, or between 4 and 4′, when these are so different in content? They have the same logical structure. To express the common logical structure of compound propositions that are different in content, we do not pay attention to the specific content of the elementary propositions constituting them and we replace these elementary propositions by variables.

This means in effect that we omit what is different in the pairs

1 and 1', 3 and 3', and 4 and 4' as they were written above and keep only that which they have in common.

We obtain forms with variables:

1-1'   If X and Y, then Z.

3-3'   If X and not Z, then not Y.

4-4'   If not Z, then not X or not Y.

In this way of writing these propositions, X, Y, and Z are variables for propositions, that is, variables for which we can substitute arbitrary propositions and not numbers as in the case of numerical variables.

Thus, the letters X, Y, and Z are used here not as the names of *specific* propositions (like the letters A, B, and C on p. 9) but as *variables* for propositions.

The pairs 1-1', 3-3', and 4-4' containing the variables X, Y, and Z are not propositions but *forms* for propositions that become propositions when we replace these variables with specific propositions. Here, the truth or falsity of the compound propositions thus obtained depends on the truth or falsity of the propositions substituted into them rather than on these elementary propositions themselves, and it is independent of their content. Therefore, it is perfectly natural to treat as the values of the variables X, Y, and Z not the specific propositions of varying content that can be substituted for these variables but instead the truth or falsity of these propositions. Thus, the values of the variables for propositions X, Y, and Z (also called **propositional variables**) are *true* and *false*.

Instead of showing individually for propositions 1 and 3 (or 1 and 4) or for propositions 1' and 3' (or 1' and 4') that the truth of each of these follows from the truth of the other, it is obviously expedient to show this just once for arbitrary propositions of the same logical structure: The fact that one does follow from the other is determined by the logical structure and not by the content of the individual propositions.

But to determine for what values of the variables the *propositional* forms 1–1′, 3–3′, and 4–4′ become true propositions, it is necessary first of all to clarify the exact meaning and properties of the logical connectives (or operations), expressed in English by the words *not, and, or, (if..., then...)*, and *if and only if*, with the aid of which various compound propositions are constructed from elementary propositions.

These operations on propositions constitute the subject of the most elementary portion of mathematical logic, known as **propositional logic** or the **algebra of propositions**. We shall use these two terms as synonyms, applying them to the same branch of logic from different points of view, namely, logic (in subject) and algebra (in method).

The first two chapters of this book are devoted to the fundamentals of propositional logic and certain of its applications. In Chapter 1, this logic is expounded as a specific meaningful theory describing a portion of the ordinary logic of human thought. In Chapter 2, the same logical system is constructed as an abstract axiomatic theory.

To distinguish between these two constructions in our terminology, we shall keep for the first the name "propositional logic" (or "algebra of propositions") and for the second we shall use the term "propositional calculus."

6. Propositional logic describes only those logical inferences for which one does not use the internal logic structure of the elementary propositions.

The investigation of the internal logical structure of the elementary propositions and their subsequent decomposition into constituent parts lead to a considerable expansion of the logical system. Propositional logic proves to be only a part of this expanded system, known as **predicate logic**. Predicate logic, like propositional logic, can be constructed in a meaningful interpretation describing a portion of the ordinary logic of human thought (still only a portion, although a greater portion than that described by propositional logic), and also, independently of any interpretation, as a purely abstract axiomatically constructed calculus.

The first construction is called **predicate logic**; the second is called the **predicate calculus** (or **functional calculus**).

Chapter 3 is devoted to an exposition of the fundamentals of predicate logic. This chapter also indicates one of the possible ways of constructing a predicate calculus.

7. Throughout our exposition, we shall make extensive use of the concept of a function.

Let M denote a set of elements of an arbitrary nature and let N also denote a set of elements of an arbitrary nature (the same as M or different). Suppose that a correspondence is set up such that for every element in M some particular element of N is put in correspondence with that element of M. Then, we say that a function f assuming values in the set N is defined on the set M. The set M is called the **domain of definition** of this function f, and the set of those elements of N that are put in correspondence with elements of the set M is called the **range of values** of the function f.

We also say that a **mapping** f of the set M is defined *into* the set N if the domain of f is the set M. The range of f need not include all values of the set N (although it may). We say that it is a mapping of the set M *onto* the set N if M is the domain and N is the range of values of f.

A function f with domain of definition M and assuming values in N (or a mapping f of the set M *into* the set N) is denoted with the aid of variables for the elements of the sets M and N as follows:

$$x \xrightarrow{f} y; x \in M, y \in N$$

or

$$x \to f(x); x \in M, f(x) \in N,$$

where the symbol $f(x)$ is a variable whose values are the values of the function f; that is, when the variable x (known as the **argument**) assumes a value a in M, $f(x)$ assumes the corresponding value $f(a)$ of the function f, where $f(a) \in N$. To indicate that f is a mapping of the set M *into* the set N, we shall use the notation:

$$M \xrightarrow{f} N.$$

When a numerical function is given with the aid of the formula (equation) $y = f(x)$, the variable $f(x)$ is expressed with the aid of a form for a number, that is, a form containing a numerical variable x. For example, in the formula $y = x^2 + 1$, $f(x)$ is expressed by the numerical form $x^2 + 1$.

A function f with domain of definition M and range of values N, that is, a mapping of the set M *onto* the set N, is denoted by the symbol

$M \xrightarrow{f} N$.

We often encounter functions whose domain of definition and range of values coincide: $M \xrightarrow{f} M$ (so that f is a mapping of the set M onto itself).

For example,

$x \xrightarrow{f} 2x + 3; x, 2x + 3 \in R$,

where R is the set of rational numbers, is a mapping of the set R onto the set R: $R \xrightarrow{f} R$.

Let us look at some examples of functions for which the domains of definition and ranges of values are finite sets. We shall encounter such functions later (Chapter 1).

a. Let M denote the set $\{a, b\}$ consisting of two elements a and b. (The nature of a and b is of no concern to us.)

Let us define the function f_1: $M \xrightarrow{f_1} M$, that is, the function with domain of definition M and range of values M, as follows:

$$\begin{bmatrix} a \xrightarrow{f_1} b \\ b \rightarrow a \end{bmatrix}$$

This table should be understood as follows: under the mapping f_1, corresponding to the element a is the element b and corresponding to the element b is the element a.

By using a variable x with range of values M: $x \in M$, we can denote this function as follows:

$x \rightarrow f_1(x); x, f_1(x) \in M$

and then write the following table, which defines for each value of the argument x the corresponding value of the function f_1:

x	$f_1(x)$
a	b
b	a

We can define yet another mapping f_2 of the set M onto itself: $M \overset{f_2}{\twoheadrightarrow} M$:

$$\begin{bmatrix} a \overset{f_2}{\rightarrow} a \\ b \rightarrow b \end{bmatrix}$$

The function f_2 puts in correspondence with the element a the element a itself and similarly puts in correspondence with the element b the element b itself. By using variables, we can indicate this function as follows:

$x \rightarrow f_2(x)$; $x, f_2(x) \in M$:

x	$f_2(x)$
a	a
b	b

There are two other functions which are mappings of the set M onto its subsets $\{a\}$ and $\{b\}$, that is, the sets consisting of only a single element a or b. These are defined by $f_3(a) = a$, $f_3(b) = a$ and $f_4(a) = b$, $f_4(b) = b$.

All four of these functions are functions of a single variable (a single argument).

b. Let us construct the set of all possible pairs of elements of the set $M = \{a, b\}$.* To do this, we take the element a as the first element of the pair and we assign to it as the second element each of the elements of M. Then, we take b as the first element of

---

* Here and throughout the remainder of the book, when we use the terms "pair," "triple," ..., "n-tuple" of elements of some set, we mean an *ordered* pair, triple, ..., set of n elements taken from the given set, allowing repetition of elements.

the pair and repeat the process. This method of constructing all possible pairs is represented by the following diagram:

$$a \Big\langle \begin{matrix} a & (a, a) \\ b & (a, b) \end{matrix}$$

$$b \Big\langle \begin{matrix} a & (b, a) \\ b & (b, b) \end{matrix}$$

Let us denote the set of all possible pairs of elements of M by M^2:

$$M^2 = \{(a, a), (a, b), (b, a), (b, b)\}.$$

We define a function φ_1 with domain of definition M^2 and range of values M; that is,

$$M^2 \xrightarrow{\varphi_1} M$$

with the aid of the following table:

$$\begin{bmatrix} (a, a) \xrightarrow{\varphi_1} a \\ (a, b) \to a \\ (b, a) \to a \\ (b, b) \to b \end{bmatrix}$$

This table defines a function φ_1 of two variables (two arguments) x and y, the range of values of these variables being in both cases the set M: $x \in M$ and $y \in M$. The range of values of the function is also the set M. This can be written

$$(x, y) \to \varphi_1(x, y); x, y, \varphi_1(x, y) \in M.$$

A table defining this function can be written as follows:

x	y	$\varphi_1(x, y)$
a	a	a
a	b	a
b	a	a
b	b	b

(If, for example, a is the number 0 and b is the number 1, so that $M = \{0, 1\}$, then the correspondence φ_1 can be interpreted as the rule assigning to each pair of elements of M the product of those elements.)

A second example is the function φ_2 ($M^2 \stackrel{\varphi_2}{\twoheadrightarrow} M$) defined by the table

$$\begin{bmatrix} (a, a) \stackrel{\varphi_2}{\to} a \\ (a, b) \to b \\ (b, a) \to b \\ (b, b) \to a \end{bmatrix}$$

The reader is invited to give a specific interpretation of this function and to show how many distinct functions $M^2 \stackrel{\varphi}{\twoheadrightarrow} M$ there are. How many distinct functions $M^2 \stackrel{\varphi}{\twoheadrightarrow} N$ are there when $N \subseteq M$ (that is, N is a subset of M, possibly, as a special case, coinciding with M)?

c. Let us consider the set of all possible triples of elements of $M = \{a, b\}$. We denote this set by M^3:

$M^3 = \{(a, a, a), (a, a, b), (a, b, a), (a, b, b), (b, a, a), (b, a, b),$
$(b, b, a), (b, b, b)\}$.

The method of forming all elements of the set M^3 is represented by the following diagram:

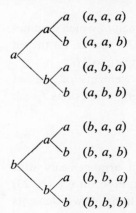

Let us now define a function ψ_1 of the three variables x, y, and z, which range over the set M: $x, y, z \in M$. The range of values of the function is also the set M: $\psi_1(x, y, z) \in M$.

This should be understood as follows: to every triple of values of the variables taken from their range of values, there is assigned a particular element in the range of values of the function. Thus, for example, this function can be defined by the following table:

x	y	z	$\psi_1(x, y, z)$
a	a	a	a
a	a	b	b
a	b	a	b
a	b	b	a
b	a	a	b
b	a	b	a
b	b	a	a
b	b	b	b

Thus, the function ψ_1 is defined on the set M^3 and has as its range of values M: $M^3 \overset{\psi_1}{\twoheadrightarrow} M$.

The reader is invited to calculate how many distinct functions ψ there are defining mappings of the set M^3 onto M.

d. Just as we defined M^2 and M^3, we can define the sets M^4, M^5, and, in general, M^n, where n is any specific natural number and $M = \{a, b\}$.

A function f of n variables $x_1, x_2, \ldots, x_n$, each of which assumes values in $M = \{a, b\}$, is defined on the set M^n, that is, on the set of all possible ordered sets of n elements belonging to M.

It is easy to determine the number of elements of the set $\{a, b\}^n$ and to see how we can construct different functions f such that

$$\{a, b\}^n \overset{f}{\twoheadrightarrow} \{a, b\}.$$

(Elements of the set $\{a, b\}^n$ are also called **arrangements with repetitions** of two elements over n positions. Procedure c (p. 23) for constructing all arrangements with repetitions of two elements over three positions indicates the procedure for determining the number of all possible arrangements with repetitions of two elements over n positions, that is, the number of elements of the set $\{a, b\}^n$.)

e. In Subsection 5, p. 7, we gave examples of various forms (expressions with variables), namely:

1. A form for a number with numerical variables;
2. A form for a proposition with numerical variables; and
3. A form for a proposition with propositional variables.

All these forms express functions.

Form 1: For example, the form for the number $x + y$, with $x, y \in N$ (where N is the set of natural numbers) defines on the set N^2 a function **S**:

$(x, y) \rightarrow S(x, y); x, y, S(x, y) \in N,$

that assumes values in N.

The function S is defined as the operation of addition in the set N: to every pair of natural numbers is assigned their sum:

$(1, 1) \rightarrow 2$
$(1, 2) \rightarrow 3$
$(2, 1) \rightarrow 3$
$(2, 2) \rightarrow 4$
$\vdots$

This table is infinite, but for an arbitrary pair of natural numbers we can determine the natural number corresponding to them (their sum) since we know the rule for adding any two natural numbers.

Thus, a form for a number with numerical variables defines a numerical function of numerical arguments.

Form 2: Let us look at the propositional form with numerical variable $x^2 + 2 = 3x$ (an equation), with $x \in C$, where C is the set of integers. In Subsection 5, we saw that, with the aid of this form, we can set up the following correspondence:

0 → F (false proposition)

1 → T (true proposition)

2 → T

3 → F

4 → F

⋮

Thus, the form for the proposition $x^2 + 2 = 3x$ containing the numerical variable x defines a function L mapping the set C onto the set $\{T, F\}$:

$C \xrightarrow{L} \{T, F\}$.

Here, as we can see, the range of values of the function is not a numerical set. It consists of the two elements T (for true) and F (for false).

Such a function assuming values in the set $\{T, F\}$ is called a **logical function**.

The form $x^2 + 2 = 3x$, where $x \in C$, expresses a logical function of a numerical variable.

Such functions (that is, logical functions of nonlogical arguments) are studied in predicate logic (see Chapter 3).

Form 3: In Section 5, we considered the propositional form

If X and Y, then Z

with propositional variables X, Y, and Z; that is,

$X, Y, Z \in \{T, F\}$.

This form itself assumes the value T or the value F when we replace the variables in it with the specific propositions that they represent, that is, their values T or F. In other words, it becomes a true or a false proposition. Consequently, it defines a function f that assigns to each triple of values of the variables X, Y, and Z in $\{T, F\}$ one of the elements of this same set $\{T, F\}$, that is, a function

$\{T, F\}^3 \xrightarrow{f} \{T, F\}$.

As we can see, the domain of definition and the range of values of this function are nonnumerical sets. These sets consist of the logical objects T and F. We have here an example of a logical function of logical arguments.

Such functions, known as **propositional functions**, are studied in propositional logic (Chapter 1).

1
PROPOSITIONAL LOGIC

1. OBJECTS AND OPERATIONS

1.1 The objects of consideration in classical logic are propositions, each of which can be true or false but which cannot simultaneously be both true and false.

Examples of propositions are the following:

a. $1 < 3$;
b. the triangle ABC is isosceles;*
c. A. S. Pushkin is the composer of the opera *Yevgeniy Onegin*;
d. $5 < 3$;
e. 3 is a prime;
f. 5 is an integral multiple of 2;
g. Nick is an outstanding pupil.*

Propositions a and e are true, whereas propositions c, d, and f are false.

If the triangle ABC referred to in Proposition b does indeed possess the property of being isosceles, i.e., if two of its sides are equal, then this proposition is true; otherwise, it is false. And since any one triangle ABC cannot both possess and fail to possess the same property, this proposition cannot be both true and false.

If the Nick referred to in Proposition g is indeed an outstanding pupil, this proposition is true, otherwise, it is false. (Of course, this proposition, like the others, cannot be both true and false simultaneously.)

Propositions of the types a–g are considered in the framework of propositional logic as **elementary**† propositions; that is, they cannot be decomposed into other propositions.

Examples of **compound**† propositions (propositions that can be decomposed into other propositions) are the following:

a′. This point lies on the line a and (this point lies)‡ on the line b.

---

* Of course, propositions b and g are only *forms* for propositions since, in both cases, the subjects are "indefinite names" (there are infinitely many triangles and many different Nicks). Here, it will be more convenient for us to assume that, in both cases, one is considering some *particular* triangle or Nick.
† See p. 15 regarding simple and compound propositions.
‡ Words usually omitted are put in parentheses.

b′. A given quadrilateral is a rhombus or (the given quadrilateral is) a rectangle.
c′. A given number is rational or (the given number is) irrational.
d′. Either he has found the solution or he has not found the solution.
e′. If a triangle is equilateral, it is isosceles.
f′. If people are intelligent and they are aware of the danger of war, then they struggle for peace.
g′. $ab = 0$ if and only if either $a = 0$ or $b = 0$.

Proposition a′ consists of two elementary propositions, namely, "this point lies on the line a" and "this point lies on the line b" connected by the word *and*.

Proposition b′ is composed of the elementary propositions "the given quadrilateral is a rhombus" and "the given quadrilateral is a rectangle" connected by the word *or*.

Proposition c′ is composed of the elementary proposition "the given number is rational" and the proposition obtained by adding the prefix "ir-"; the two are connected by the word *or*.

Proposition d′ is composed of the elementary proposition "he has found the solution" and the proposition obtained by inserting the word *not*; they are connected by the words *either ...or*.

Proposition e′ is composed of the elementary propositions "the triangle is equilateral" and "the triangle is isosceles"; connected by the words *if..., then*.

Proposition f′ is composed of the elementary propositions "people are intelligent," "they are aware of the danger of war," and "they struggle for peace." The three are connected in a special way by the words *if..., then* and *and*.

Proposition g′ is composed of the propositions "$ab = 0$," "$a = 0$," "$b = 0$"; they are connected in a special way by the words *if and only if* and *or*.*

---

* Since the truth or falsity of a compound proposition is, as we shall show, uniquely determined by the truth or falsity of the elementary propositions comprising it, it follows that a compound proposition can, like an elementary proposition, be true or false but cannot be both true and false at the same time.

The words *not, and, or, (if..., then), if and only if,* and a few others indicate in ordinary language *logical operations* with the aid of which new propositions are formed out of old ones (elementary or compound).

These operations on propositions constitute the subject matter of the algebra of propositions. In what follows, we shall denote propositions by capital italic letters with or without a subscript:

$A, B, C, \ldots, X, Y, Z; X_1, X_2, \ldots, X_n.$

These letters can also be used both as names of specific propositions and as propositional variables in place of which one can substitute arbitrary propositions, just as letters are used in elementary algebra both as the names of numbers and as variables for numbers.

The values of propositional variables are T (for true) and F (for false). T and F are called the **truth values**.*

The truth values are sometimes denoted by symbols other than T or F. Thus, for example, in certain applications, it is convenient to denote T with 1 and F with 0 (or the other way around) without, of course, ascribing to the symbols 1 and 0 any other "meaning" than truth values when we do this.

EXERCISES

1.01. In the following propositions, pick out the components, denote them by letters, and underscore the logical connectives:

a. If the diagonals of a parallelogram are perpendicular to each other, that parallelogram is a rhombus.

b. $a \geq 0$.

c. A given number either is an integral multiple of 2 and an integral multiple of 3 or it is not an integral multiple of 6.

d. If a point A belongs to a straight line a and the straight line a belongs to a plane α, then the point A belongs to the plane α.

1.02. In the following compound propositions, denote the ele-

---

* Instead of saying "truth value," we shall frequently say simply "value" (just as, in elementary algebra, we speak of the "value" instead of the "numerical value").

mentary propositions by letters* and determine which of these compound propositions have the same logical structure.

a. If the diagonals of a parallelogram are perpendicular to each other or if they bisect its angles, that parallelogram is a rhombus.

b. If one of two numbers is an integral multiple of 3 and if the sum of the two numbers is an integral multiple of 3, then the other number is an integral number of 3.

c. If a parallelogram is not a rhombus, its diagonals are not perpendicular to each other and they do not bisect the angles of the parallelogram.

d. If $a = 0$ or $b = 0$, then $ab = 0$.

e. If $ab \neq 0$, then $a \neq 0$ and $b \neq 0$.

f. If $a > 0$ and $b > 0$, then $ab > 0$.

1.2. Logical operations expressed in everyday language by the words or phrases *not*, *and*, *or*, (*if...*, *then*), *if and only if* transform propositions into other propositions. For each of these operations, the operation itself and the result of its execution are denoted by the same term.

For example, the operation expressed by the adverb *not* is called **negation** and the proposition obtained by applying it to a given proposition is also called the **negation** (of the given proposition). (This operation, unlike the others that we shall consider below, is applied to a single proposition.)

The operation expressed by the connective *and* is itself called **conjunction** and the proposition obtained by applying this operation to two given propositions is also called a **conjunction** (of these propositions), etc. (This terminology differs from the terminology of ordinary algebra, where the operation and the result of its application are indicated by different terms such as "addition" but "sum," or "multiplication" but "product." It might be noted that sometimes such terminology is also applied in the algebra of propositions.)

---

* A single elementary proposition should be denoted by the same letter whenever it occurs in compound propositions a–f.

In the following operations, in order to shorten the formulations we shall define not the operations themselves but the results of their application; that is, by "negation," "conjunction," etc., we shall mean in these definitions the results of applying the operations in question. In giving definitions of the logical operations, we shall try as far as possible to have these definitions correspond to the everyday meaning of the words *not*, *and*, *or*, (*if...*, *then*), and *if and only if*.

a. *Negation.* The negation of a given proposition is defined as the proposition that is true if the given proposition is false and false if the given proposition is true.

We denote the negation of a proposition X by $\overline{X}$. For example, the negation of the proposition $A \in a$ is denoted by $\overline{A \in a}$, the negation of the proposition $a \parallel b$ is denoted by $\overline{a \parallel b}$ and the negation of $1 < 3$ is denoted by $\overline{1 < 3}$.*

The negation of an operation can be written in the form of the following table:

X	$\overline{X}$
T	F
F	T

This table, which is called a **truth table,** shows how the truth value (T or F) of the negation $\overline{X}$ depends upon the truth value of the proposition X.

This table defines a propositional function of a single variable X, namely, the mapping of the set {T, F} onto itself that assigns to the element T the element F and to the element F the element T.

One can easily see that the proposition X is the negation of the proposition $\overline{X}$ (since it satisfies the definition of negation).

b. *Conjunction.* If two propositions are connected by the connective *and*, the resulting proposition is usually assumed to be true if and only if each of the propositions comprising it is true. If at least one of these component propositions is false, then the

* Other notations are ⌐ and ~ (placed in *front* of the proposition being negated).

34 CH. 1. PROPOSITIONAL LOGIC

compound proposition obtained by connecting them by means of the word *and* is also false.

For example, the proposition "the number 2 is prime and even" is true since both propositions comprising it, namely, "the number 2 is prime" and "the number 2 is even" are true. The proposition "the diagonals of an isosceles trapezoid are equal and are bisected by the point of intersection" is false since the second of the two propositions comprising it, namely, "(the diagonals of an isosceles trapezoid) are bisected by the point of intersection" is false.

The conjunction* of two propositions X and Y is defined as that proposition that is true if and only if the propositions X and Y are both true. We denote this proposition by $X \wedge Y$ (the symbol $\wedge$ replaces the word *and*).†

The definition of conjunction can be written in the form of the following truth table: ‡

X	Y	$X \wedge Y$
T	T	T
T	F	F
F	T	F
F	F	F

This table defines a propositional function of two variables X and Y on the set $\{T, F\}^2$ with the set $\{T, F\}$ as its range of values.

The definition of conjunction can be extended to an arbitrary number of propositions. The conjunction $X_1 \wedge X_2 \wedge \cdots \wedge X_n$, which we usually denote by $\bigwedge_{i=1}^{n} X_i$ is true if and only if each of the

---

* Called also the (logical) **product**.

† A conjunction is also denoted sometimes by the symbol & or by a dot, the symbol for ordinary multiplication, which, just as in ordinary algebra, is often omitted (XY instead of $X \cdot Y$).

‡ If X and Y are definite propositions (that is, if they are the names of propositions), then $X \wedge Y$ is also a definite proposition. On the other hand, if either X or Y (or both) contains a propositional variable, then $X \wedge Y$ is a propositional form.

propositions $X_1, X_2, \ldots, X_n$ is true (and it is false if at least one of these propositions is false).

c. *Disjunction.* In ordinary speech, the word *or* is applied in two distinct senses: a nonexclusive sense when a compound proposition formed with its aid is considered true if *at least* one of its component propositions is true and an exclusive sense when the compound proposition is considered true if *one* of the component propositions is true but not both (in this case we sometimes say *either...or*).

For example, if in the proposition "this pupil is capable or industrious" the word *or* is applied in the nonexclusive sense, this proposition is true if at least one of the propositions comprising it is true, that is, if the pupil in question is capable or industrious or both capable and industrious. It is false only if both these component propositions are false, that is, if the pupil is both incapable and lazy.

If in the proposition "Andrew will become a physicist or a mathematician" the word *or* is understood in the other (exclusive) sense, then this proposition is considered true if one of the component propositions is true and the other false, that is, if Andrew either will become a physicist but not a mathematician or he will become a mathematician but not a physicist. It is considered false if both component propositions are false or both true, that is, if Andrew will become neither a physicist nor a mathematician or he will become both. (In the latter case, we usually say "we guessed wrongly in predicting that Andrew would become a physicist *or* a mathematician: he became both.")

In everyday language, it is in general difficult to distinguish in which sense the word *or* is used in a proposition. In the language of present-day logic, this ambiguity is eliminated by using a distinct notation for the two meanings of the word *or*.

Beginning with the nonexclusive sense of the word *or*, we have the following definition:

The **disjunction*** of two propositions is defined as the new

---

* Also called the (logical) **sum**.

proposition that is true if and only if *at least* one of these propositions is true.

The disjunction of two propositions X and Y is denoted by $X \vee Y$ (that is, the symbol $\vee$ takes the place of the word *or* in its nonexclusive sense).

The definition of disjunction can be written in the form of the following truth table:

X	Y	$X \vee Y$
T	T	T
T	F	T
F	T	T
F	F	F

This table like the conjunction table defines a function* of the two variables X and Y on the set $\{T, F\}^2$, the range of which is the set $\{T, F\}$.

The definition of disjunction, like the definition of conjunction, can be extended in a natural manner to an arbitrary number of component propositions.

EXERCISES

1.03. When is the disjunction $\bigvee_{i=1}^{n} X_i$ true? When is it false? When is the conjunction $\bigwedge_{i=1}^{n} X_i$ false?

1.04. What are the domains of definition and ranges of the functions defined by the disjunction $\bigvee_{i=1}^{n} X_i$ and the conjunction $\bigwedge_{i=1}^{n} X_i$? How many rows does the table have corresponding to each of these functions?

1.05. The operation corresponding to the connective *or* in the exclusive sense is called **strict disjunction**. Define the strict

---

* Since we are not considering any functions other than propositional functions in the present book, we shall say simply "function" instead of "propositional function."

disjunction of two propositions X and Y and set up the corresponding table. (Strict disjunction can be denoted by the symbol $\dot{\vee}$, that is, the symbol for ordinary disjunction with a dot over it.)

1.06. Set up truth tables for the propositions

a. $X \wedge Y \wedge Z$;

b. $X \vee Y \vee Z$.

    d. *Implication.* Quite frequently (and not only in mathematics), we use compound propositions composed of two propositions connected by the words *if...*, *then*.

A compound proposition of the form

If X, then Y.

called a **conditional proposition**, is a proposition about entailment (or inference). X is the **premise** and Y is the **consequence** or **conclusion**.

In everyday speech, we have numerous synonyms for "if X, then Y":

Y follows from X,

X implies Y,

X is a sufficient condition for Y,

Y if X,

Y is a necessary consequence of X,

Y under the condition that X,

and so forth.

The conditional proposition "if X, then Y," that is, the proposition that Y follows from X, is characterized by the following two conditions:

1. It is false if and only if the premise (X) is true and the conclusion (Y) is false,
2. X and Y are related in content, in meaning.

However, in formal deductions, we pay no attention to the content of the premise and conclusion and hence no attention to their relationship as regards content but consider only the relationship between their truth values, that is, the satisfaction of Condition 1. Therefore, in the logic of propositions, we usually neglect Condition 2, and, in defining the operation corresponding to the operation of obtaining a conditional proposition, we consider only Condition 1.

This operation (and the result of its application) we call **implication**.

We denote the implication the premise of which is X and the conclusion of which is Y by writing*

$X \Rightarrow Y$,

which we read "X implies Y."

To define implication, we use a logical characteristic of a conditional proposition (that is, the fact that it is false if and only if the premise is true and the conclusion false) and arrive at the following definition:

The implication $X \Rightarrow Y$ is that proposition which is false if and only if X is true and Y false.

This definition can be written in the form of the following truth table:

X	Y	$X \Rightarrow Y$
T	T	T
T	F	F
F	T	T
F	F	T

Of course, neglect of Condition 2 leads to such "unnatural" consequences as our accepting as true not only implications of the forms

a. If a (some particular number) is an integral multiple of 3, then a^2 is also an integral multiple of 3,

* Other symbols used for this are $\rightarrow$ and $\supset$.

b. If 9 is an integral multiple of 3, then 81 is also an integral multiple of 3,

c. If 10 is an integral multiple of 3, then 100 is also an integral multiple of 3;

but even

d. If $2 \times 2 = 4$, then Paris is the capital of France,
e. If $2 \times 2 = 5$, then Paris is the capital of France,

and

f. If $2 \times 2 = 5$, then London is the capital of France.

Of course, the truth of b and c (even though the first of these implications appears trivial and the second unnecessary) follows immediately from comparison with a, of which they are special cases. We are familiar with this type of specific consequence of general propositions from the proofs of mathematical theorems, and their only difference from "ordinary" everyday phraseology is that in mathematics we do not usually use the subjunctive mood, considering it a purely superfluous encumbrance. (Indeed, the implication c′, "if 10 *were* an integral multiple of 3, then 100 *would be* an integral multiple of 3," *would* entail no objections or confusion.)

With regard to d, e, and f, we have the "paradoxical" consequences of dispensing with Condition 2; however, such a course is necessary. The point is that in defining logical operations (not just the operation of implication), we are governed only by the following condition, which is characteristic and decisive for the *formalization* of logic:

3. The truth or falsity of the conclusion is *uniquely* determined by the truth or falsity of the premises.

Condition 3 was satisfied in the definition of the operations of negation, conjunction, and disjunction (and, in fact, it was its strict satisfaction that led us to the necessity of our remarks concerning the different meanings of the word *or*). Application of

Condition 3 to the question of the definition of an implication can be represented in the form of a process of step-by-step compilation of a truth table for an implication:

	X	Y	$X \Rightarrow Y$
(1)	T	T	
(2)	T	F	
(3)	F	T	
(4)	F	F	

1–2: The naturalness of the requirement that a true premise leads under "correct" reasoning without fail to a true and not a false conclusion leads to the fact that the last position in the first row of the table is filled with T and that the last position in the second row is filled with the value F.

3–4: Since we have accepted a false premise, we can, by *correct* reasoning, arrive either at a true conclusion (Case 3, Example e) or a false conclusion (Case 4, examples b and f); the third and fourth rows must be filled with the value T.

Of course, despite the truth of implication f, no "entailment" is expressed in it since the premise and the conclusion are unrelated in content. In ordinary speech, such a proposition would be considered senseless.

This phenomenon, the existence of true implications that are not propositions about entailment, is, of course, explained by the fact that, in our definition of an implication, we did not observe Condition 2, which assumes a relationship in content between the premise and the conclusion. Therefore, the class of implications that is defined with the aid of Condition 1 is broader than the class of conditional propositions that satisfy both conditions 1 and 2 and contains the latter class as a subclass: every true conditional proposition is expressed by a true implication but not every true implication is the expression of a conditional proposition in the usual sense.

In mathematical logic, logical systems are developed with yet other forms of implications, in particular, with the purpose of reflecting certain aspects of the relationship "in meaning" between the premise and the conclusion.*

e. *Equivalence.* The compound proposition "a quadrilateral is a parallelogram *if and only if* its diagonals are bisected by their point of intersection" is considered true if its components, that is, the proposition "the quadrilateral is a parallelogram" and the proposition "its diagonals are bisected by their point of intersection," are both true or both false.

A compound proposition consisting of two propositions X and Y connected by the phrase *if and only if* is called an **equivalence** and we denote it by $X \Leftrightarrow Y$, where the symbol $\Leftrightarrow$ stands for the words *if and only if*.†

Thus, beginning with the ordinary meaning of the phrase *if and only if*, we arrive at the following definition:

The equivalence of two propositions X and Y is defined as the proposition that is true if and only if both these propositions X and Y are true or else both are false.

This definition can be written in the form of the following truth table:

X	Y	$X \Leftrightarrow Y$
T	T	T
T	F	F
F	T	F
F	F	T

The truth of the composite proposition is usually established by proving the two theorems (converses of each other) "*if* a quadrilateral is a parallelogram, *then* its diagonals are bisected by their point of intersection" and "*if* the diagonals of a quadrilateral are bisected by their point of intersection, *then* the quadrilateral is a parallelogram."

---

* See Lewis, C. I. and C. H. Langford, *Symbolic Logic*, New York and London, 1932 (and reprinted in 1959 by Dover, New York); and Curry, H. B., *Foundations of Mathematical Logic*, McGraw-Hill, 1963.

† Other symbols in use are $\leftrightarrow$, $\sim$, and $\equiv$.

Let X denote the proposition "a quadrilateral is a parallelogram" and let Y denote the proposition "the diagonals of the quadrilateral are bisected by their points of intersection."

Then, our compound proposition may be written

If X, then Y, and if Y, then X.

If we replace the words *if...*, *then* with implication symbols and the grammatical conjunction *and* with the conjunction symbol, we obtain

$$(X \Rightarrow Y) \wedge (Y \Rightarrow X). \tag{1}$$

This compound proposition is constructed from elementary propositions by using the operations of implication and conjunction with which we are now familiar, and we can construct for it a truth table analogous to those constructed above, beginning with the definitions (tables) of these operations.

In the first two columns, we write all possible combinations of values of the two elementary propositions X and Y:

1	2	3	4	5
X	Y	$X \Rightarrow Y$	$Y \Rightarrow X$	$(X \Rightarrow Y) \wedge (Y \Rightarrow X)$
T	T	T	T	T
T	F	F	T	F
F	T	T	F	F
F	F	T	T	T

In the third column, we write the values of the implication $X \Rightarrow Y$ corresponding to these sets of values of X and Y in accordance with the definition of an implication. In the fourth column, we write analogously the values of the implication $Y \Rightarrow X$. Finally, in the fifth column we write the values of Conjunction 1 corresponding to the values of the implications $X \Rightarrow Y$ and $Y \Rightarrow X$ that we have written in columns 3 and 4.

The table that we have obtained (its first, second, and fifth columns) coincides completely with the table given above for the operation of equivalence, which can thus be reduced to implication and conjunction.

We can also consider $X \Leftrightarrow Y$ as simply a short way of writing Expression 1. The symbol $\Leftrightarrow$ suggests the double implication $\genfrac{}{}{0pt}{}{\Rightarrow}{\Leftarrow}$.

1.3. The definitions given above for the logical operations enable us to find the truth value of a compound proposition if we know the truth values of the elementary propositions from which it is constructed with the aid of these operations.

For example, suppose that we are given the compound proposition

$$((X \vee Y) \Rightarrow (\bar{X} \wedge \bar{Z})) \Leftrightarrow (Y \vee Z)$$

and we know that X, Y, and Z have the values T, T, and F, respectively.

The determination of the value of the compound proposition can be written in the form of the following scheme (which obviously does not need comment):

$$((X \vee Y) \Rightarrow (\bar{X} \wedge \bar{Z})) \Leftrightarrow (Y \vee Z)$$

```
((X ∨  Y) ⇒ (X̄ ∧  Z̄)) ⇔ (Y ∨  Z)
   T   T      F    T        T   F
    ⏟         ⏟            ⏟
    T         F            T
         ⏟                     
         F                 T
              ⏟
              F
```

Consequently, the given compound proposition has the value F.

EXERCISES

1.07. From the elementary propositions

A, this number is an integer,
B, this number is positive,
C, this number is a prime,
D, this number is an integral multiple of 3,

we construct the following composite propositions:
a. $A \vee B$
b. $A \wedge B$
c. $A \vee \bar{A}$
d. $B \wedge \bar{B}$
e. $D \Leftrightarrow \bar{C}$
f. $(A \wedge C) \Rightarrow \bar{D}$
g. $(A \wedge D) \Rightarrow \bar{C}$
h. $(A \vee B) \wedge (C \vee D)$
i. $\bar{A} \vee \bar{D}$
j. $(A \wedge B \wedge C) \vee D$

Read all these propositions in nonmathematical English.

1.08. Construct a table for $\bar{X} \vee Y$ and compare it with the table for the implication $X \Rightarrow Y$. What conclusion can be drawn from this comparison?

1.09. a. What are synonyms for the proposition "X if and only if Y"? b. Write the proposition "a number is rational if and only if it can be represented in the form of the ratio of two integers" as a conjunction of two implications.

1.10. Determine the values of the following compound propositions:
a. $X \wedge (Y \wedge Z)$
b. $(X \wedge Y) \wedge Z$
c. $X \Rightarrow (Y \Rightarrow Z)$
d. $(X \wedge Y) \Rightarrow Z$
e. $(X \wedge Y) \Leftrightarrow (Z \vee \bar{Y})$
f. $[(X \vee Y) \wedge Z] \Leftrightarrow [(X \wedge Z) \vee (Y \wedge Z)]$

if we know that X has the value F and that Y and Z both have the value T.

1.11. a. Suppose that we know that the implication $X \Rightarrow Y$ is true and that the equivalence $X \Leftrightarrow Y$ is false. What can we say about the value of the implication $Y \Rightarrow X$?

b. Suppose that we know that the equivalence $X \Leftrightarrow Y$ is true. What can we say about the values of $\bar{X} \Leftrightarrow Y$ and $X \Leftrightarrow \bar{Y}$?

c. Suppose that we know that X has the value T. What can we say regarding the values of the implications
$(\bar{X} \wedge Y) \Rightarrow Z$, $\bar{X} \Rightarrow (Y \vee Z)$?

d. Suppose that we know that the implication $X \Rightarrow Y$ has the value T. What can we say about the values of the implications $Z \Rightarrow (X \Rightarrow Y)$, $\overline{(X \Rightarrow Y)} \Rightarrow Y$, $(X \Rightarrow Y) \Rightarrow Z$?

1.12. Consider the compound proposition "if the base of a pyramid is a regular polygon and the altitude passes through the center of the base or the dihedral angles at the base are equal, then the pyramid is a regular pyramid." Denote the elementary propositions in this compound proposition by letters and replace the words expressing logical operations with the appropriate symbols. Where can we put parentheses so that the compound proposition is true and where so that it is false?

2. FORMULAS. EQUIVALENT FORMULAS. TAUTOLOGIES

2.1 An expression consisting of letters, symbols for the operations introduced in Section 1, and parentheses, which is used to denote a proposition or a form for a proposition, is called a **formula** in the algebra of propositions.* It is intuitively clear, for example, that $\overline{X}$, $X \wedge Y$, and $X \Rightarrow Y$ are formulas and that the expression $X \wedge \Rightarrow$ is not a formula. However, in more complicated cases, our intuition can prove insufficient.

This definition enables us to determine whether or not a certain expression is a formula only on the basis of the meaning of the symbols of which this expression is composed. In accordance with this explanation, we must also consider as formulas the individual letters $A, B, C, \ldots, X, Y, Z$, which denote propositional variables (which, in the present chapter, we shall call simply variables) and also letters denoting logical constants, that is, T (a true proposition) and F (a false proposition).

The symbols $\overline{}$, $\wedge$, $\vee$, $\Rightarrow$, and $\Leftrightarrow$, also denoting logical constants, are, of course, not formulas when taken alone.

Parentheses, just as in ordinary algebra, play the role of punctuation marks used to specify the order of operations.

To simplify our writing of formulas (that is, to reduce the number of parentheses in them), let us adopt the following conventions:

1. We shall not include in parentheses a formula or a portion of a formula that lies underneath the symbol for negation; that is, we shall write $\overline{X \vee Y} \wedge Z$ instead of $(\overline{X \vee Y}) \wedge Z$.

---

* A precise definition will be given in Chapter 2.

2. We shall assume that the symbol for conjunction connects letters more closely than do the symbols for disjunction, implication, and equivalence; that is, we shall write

$X \wedge Y \vee Z$ instead of $(X \wedge Y) \vee Z$;
$X \Rightarrow Y \wedge Z$ instead of $X \Rightarrow (Y \wedge Z)$;
$X \wedge Y \Leftrightarrow Z \wedge T$ instead of $(X \wedge Y) \Leftrightarrow (Z \wedge T)$.

3. We shall assume that the symbol for disjunction connects letters more closely than do the symbols for implication and equivalence; that is, we shall write

$X \vee Y \Rightarrow Z$ instead of $(X \vee Y) \Rightarrow Z$;
$X \Leftrightarrow Y \vee Z$ instead of $X \Leftrightarrow (Y \vee Z)$.

4. We shall assume that the symbol for implication connects letters more closely than does the symbol for equivalence; that is, we shall write

$X \Rightarrow Y \Leftrightarrow Z$ instead of $(X \Rightarrow Y) \Leftrightarrow Z$.

These conventions will considerably simplify the writing of our formulas. For example, the formula that without them would have to be written

$(((X \vee Y) \wedge Z) \Rightarrow (Z \vee \bar{X})) \Leftrightarrow ((X \vee Y) \Rightarrow ((Y \wedge Z) \vee \bar{X}))$

can now be written

$(X \vee Y) \wedge Z \Rightarrow Z \vee \bar{X} \Leftrightarrow X \vee Y \Rightarrow Y \wedge Z \vee \bar{X}$.

In reading, a formula can be called by the name of that operation that is carried out last (that is, the operation whose symbol connects more weakly than the remaining symbols of the operations appearing in the formula). Thus, the formula written above is an equivalence (of two implications).

2.2. We denote an arbitrary formula containing variables $X_1, X_2, \ldots, X_n$ by

$f(X_1, X_2, \ldots, X_n)$.

This formula defines on the set of all possible n-tuples of values of the variables a function f whose arguments as well as its values are in $\{T, F\}$.

A formula is only one of the various possible forms of defining a function f. Since every variable X_i assumes only two values (T and F), the domain of definition of the function f, that is, the set $\{T, F\}^n$, is a finite set; specifically, it contains 2^n elements. Therefore, such a function can always be defined by giving a finite truth table similar to those that we constructed in Section 1. The truth table corresponding to the function f (we shall also say that it corresponds to the formula $f(X_1, X_2, \ldots, X_n)$ defining this function) contains 2^n rows. In each row, for a particular n-tuple of values of the variables, we write the corresponding value of the function, which we shall also call the value of the formula. We have already seen (Section 1.2, Part e) how a truth table is composed for a formula containing more than one operational symbol.

EXERCISE

1.13. Set up truth tables corresponding to the following formulas:

a. $\overline{X \vee Y}$
b. $\overline{X} \wedge \overline{Y}$
c. $X \wedge (Y \vee Z)$
d. $(X \wedge Y) \vee (X \wedge Z)$
e. $X \Rightarrow Y \vee Z$
f. $(X \Rightarrow Y) \vee (X \Rightarrow Z)$
g. $X \vee (Y \wedge Z)$
h. $(X \vee Y) \wedge (X \vee Z)$

2.3. Formulas different in structure can define the same function.
For example, in Exercise 1.13, formulas a and b define a single function; so do formulas c and d, formulas e and f, and formulas g and h. This can be seen from their truth tables.

Formulas defining the same function are said to be **equivalent**.

The relation of equivalence is a relation between formulas of the algebra of propositions. A term or a symbol denoting this relation belongs not to the language of the algebra of propositions but to the language used to describe and study the formalized language of the algebra of propositions (the **metalanguage***).

---

* In some logical studies we investigate logical properties of formulas. The formulas are thought of as belonging to a *language*, sometimes called the *object* language. However, in stating the results of such investigations we need to use a

We shall denote equivalence by writing "equiv" as an abbreviation for the phrase "is equivalent to."

(This relation is sometimes denoted by the ordinary equal sign = or the identity symbol ≡, or by the symbol ∼.)

The statement "$\overline{X \vee Y}$ equiv $\overline{X} \wedge \overline{Y}$" (which means that the formulas $\overline{X \vee Y}$ and $\overline{X} \wedge \overline{Y}$ are equivalent) is not a formula in the algebra of propositions but a proposition about the formulas of that algebra that is expressed in the metalanguage (in the present case, ordinary English) in which we are describing and studying that algebra.

It follows directly from the definition of the relation of equivalence that it is reflexive, symmetric, and transitive, that is, that

1. φ equiv φ for an arbitrary formula φ (reflexivity);
2. If φ_1 equiv φ_2, then φ_2 equiv φ_1 for arbitrary formulas φ_1 and φ_2 (symmetry);
3. If φ_1 equiv φ_2 and φ_2 equiv φ_3, then φ_1 equiv φ_3 for arbitrary formulas φ_1, φ_2, and φ_3 (transitivity).

The fact that we can define a single function by means of several equivalent formulas reflects the fact that we can express a single thought with the aid of propositions of different logical structure.

Let us give a few examples of equivalent formulas.

a. The implication $X \Rightarrow Y$ assumes the value F only when X has the value T and Y has the value F. But it is for just these values of the variables that the disjunction $\overline{X} \vee Y$ assumes the value F. For all other combinations of values of the variables, these two formulas both assume the value T. Thus, the same truth table corresponds to these formulas and they define the same function; that is,

$X \Rightarrow Y$ equiv $\overline{X} \vee Y$.

---

language, and this language will in general be other than the object language; it is often called the *metalanguage*. The sign "equiv" introduced above is thus one belonging to the metalanguage. It is used to make claims *about* formulas, but does not occur in them, and thus does *not* belong to the object language. In this respect it is quite different from "⇔" which does so belong.

b. Let us return to one of the examples that we considered in Section 5 of the Introduction. There, it was asserted (at the time without any substantiation) that if the form

1–1′   if X and Y, then Z

becomes a true proposition, then each of the following two forms also becomes a true proposition:

3–3′   If X and not Z, then not Y

and

4–4′   if not Z, then not X or not Y,

and conversely.

Now, we can prove this. To do this, we need only show that forms 1–1′ and 3–3′ are expressed by equivalent formulas and that forms 1–1′ and 4–4′ are expressed by equivalent formulas.

Let us show that

$$X \wedge Y \Rightarrow Z \text{ equiv } X \wedge \bar{Z} \Rightarrow \bar{Y}$$

and

$$X \wedge Y \Rightarrow Z \text{ equiv } \bar{Z} \Rightarrow \bar{X} \vee \bar{Y},$$

that is, that the three formulas $X \wedge Y \Rightarrow Z$, $X \wedge \bar{Z} \Rightarrow \bar{Y}$, and $\bar{Z} \Rightarrow \bar{X} \vee \bar{Y}$ are all equivalent by using the following truth table:

1 2 3 X Y Z	4 5 6 7 $\bar{X}$ $\bar{Y}$ $\bar{Z}$ $X \wedge Y$	8 $X \wedge Y \Rightarrow Z$	9 $X \vee \bar{Z}$	10 $X \wedge \bar{Z} \Rightarrow \bar{Y}$	11 $\bar{X} \vee \bar{Y}$	12 $\bar{Z} \Rightarrow \bar{X} \vee \bar{Y}$
T T T	F F F T	T	F	T	F	T
T T F	F F T T	F	T	F	F	F
T F T	F T F F	T	F	T	T	T
T F F	F T T F	T	T	T	T	T
F T T	T F F F	T	F	T	T	T
F T F	T F T F	T	F	T	T	T
F F T	T T F F	T	F	T	T	T
F F F	T T T F	T	F	T	T	T

Coincidence of the values of the three formulas for any set of values of the variables (columns 8, 10, and 12) proves their equivalence.

c. The proposition "it is untrue that the point A belongs to the straight line a and the straight line b"* denotes the same thing as the proposition "the point A does not belong to the line a or it does not belong to the line b."

If we disregard the specific content of these compound propositions, replacing the elementary propositions constituting them with variables (letting X denote "the point A belongs to the line a" and letting Y denote "the point A belongs to the line b"), we obtain the formulas $\overline{X \wedge Y}$ and $\overline{X} \vee \overline{Y}$.

Beginning with the definition of the operations or using a truth table, we can easily establish the equivalence of these formulas; that is, we can prove that

$$\overline{X \wedge Y} \text{ equiv } \overline{X} \vee \overline{Y}.$$

Equivalent formulas in the algebra of propositions are the analogues of identities of ordinary algebra. However, in ordinary algebra as opposed to the algebra of propositions, we cannot prove the identity of two expressions, for example, $(a + b)^2$ and $a^2 + 2ab + b^2$, by direct verification of the coincidence of their values for all possible combinations of the values of the variables a and b, since there are infinitely many such combinations. The identity of these two expressions is shown by transforming one into the other.

The equivalence of formulas of the algebra of propositions can be established with the aid of the truth tables corresponding to these formulas, that is, essentially by direct verification of the coincidence of their values for all possible combinations of values of the variables.

However, although truth tables are always finite, when the number of variables is large the compilation of these tables is

---

* The phrase "it is untrue that" denotes negation of the entire proposition following this phrase.

inconvenient from a practical standpoint since the number of rows increases quite rapidly with increase in the number of variables.

In the algebra of propositions just as in ordinary algebra, certain pairs of equivalent formulas are used to prove the equivalence of other formulas by transforming one into the other. We present below the most important equivalences expressing the properties of the operations in the algebra of propositions:

1. $\bar{\bar{X}}$ equiv X
2. $X \vee Y$ equiv $Y \vee X$
3. $X \wedge Y$ equiv $Y \wedge X$
4. $(X \vee Y) \vee Z$ equiv $X \vee (Y \vee Z)$
5. $(X \wedge Y) \wedge Z$ equiv $X \wedge (Y \wedge Z)$
6. $X \wedge (Y \vee Z)$ equiv $(X \wedge Y) \vee (X \wedge Z)$
7. $X \vee (Y \wedge Z)$ equiv $(X \vee Y) \wedge (X \vee Z)$
8. $X \vee X$ equiv X
9. $X \wedge X$ equiv X
10. $X \vee F$ equiv X
11. $X \wedge T$ equiv X
12. $X \wedge F$ equiv F
13. $X \vee T$ equiv T
14. $X \vee \bar{X}$ equiv T
15. $X \wedge \bar{X}$ equiv F
16. $\overline{X \wedge Y}$ equiv $\bar{X} \vee \bar{Y}$
17. $\overline{X \vee Y}$ equiv $\bar{X} \wedge \bar{Y}$
18. $X \Rightarrow Y$ equiv $\bar{X} \vee Y$
19. $X \Leftrightarrow Y$ equiv $(X \Rightarrow Y) \wedge (Y \Rightarrow X)$
20. $(X \wedge Y) \vee \bar{X}$ equiv $Y \vee \bar{X}$
21. $(X \vee Y) \wedge \bar{X}$ equiv $Y \wedge \bar{X}$

All these equivalences are easily proved by starting with the definitions of the operations and using truth tables. (Some of these we have already proved, some of the others are included in the exercises.) In what follows, we shall have frequent occasion to cite these formulas; in such cases, we shall refer to them by number.

Let us look at this list carefully.

a. Relations 2–13, containing only symbols for disjunction and conjunction, show that these operations, just like addition and multiplication in ordinary algebra, are commutative and associative (2–5). However, whereas in ordinary algebra multiplication is distributive with respect to addition but not vice versa, in the algebra of propositions, each of the two operations, disjunction and conjunction, is distributive with respect to the other (6, 7). If we replace the symbol for disjunction by the symbol $+$, the symbol for conjunction by the symbol $\times$, the letter F by 0, and the letter T by 1, then the properties analogous to those expressed in relations 10–12 hold in ordinary algebra, but the analogues of the properties expressed in 8, 9, and 13 do not hold.

b. Let φ denote some formula constructed with the aid of the symbols for negation, conjunction, and disjunction. If we replace the symbols $\wedge$, $\vee$, T, and F by the symbols $\vee$, $\wedge$, F, and T, respectively, the resulting formula φ' is called the **dual** of φ.

If we apply this procedure to the formula φ', we obtain the formula φ; that is, φ is the dual of φ'. Thus, the relationship of duality is a reciprocal one.

For example, in each of the following pairs of formulas, the first formula is the dual of the second and vice versa:

$X \wedge (Y \vee Z)$, $X \vee (Y \wedge Z)$;
$\overline{X \wedge Y}$, $\overline{X \vee Y}$;
$X \wedge T$, $X \vee F$.

A careful examination of relations 2–17 shows that they can be divided into pairs of dual formulas: 2 and 3, 4 and 5, 6 and 7, etc.

It turns out that there is a *duality law* or *principle* in the algebra of propositions:

If two formulas are equivalent, their duals are also equivalent.

Below, we shall illustrate this principle with an example.*

c. Let us show that certain of the equivalences (1–21) can be obtained from others. In these transformations, we shall use two

---

* For a proof, see Appendix I.

rules analogous to those that we use in the identical transformations of the expressions of ordinary algebra.

Rule 1: A variable that appears in the composition of two equivalent formulas can be replaced everywhere that it appears in them by the same arbitrary formula. Since this formula assumes the same values (T and F) as does the variable that it replaces, this substitution yields equivalent formulas. Application of this rule enables us to obtain from 1–21 as many new equivalences as we wish.

Rule 2: A formula constituting a portion of any given formula can be replaced with a formula equivalent to it. Since this replacement does not change the value of the entire formula, we obtain as a result of it a formula equivalent to the original one.

Let us now show that we can obtain from Equivalence 6, which expresses the distributivity of conjunction with respect to disjunction, Equivalence 7, which expresses the distributivity of disjunction with respect to conjunction. (From this same example, we can see the idea of the proof of the duality principle.)

Suppose that we are given the equivalence

$X \wedge (Y \vee Z)$ equiv $(X \wedge Y) \vee (X \wedge Z)$ (Equivalence 6).

If two formulas are equivalent, then their negations are also equivalent:

$\overline{X \wedge (Y \vee Z)}$ equiv $\overline{(X \wedge Y) \vee (X \wedge Z)}$.   (a)

From a and equivalences 16 and 17, we obtain, in accordance with Rule 1,

$\overline{X} \vee \overline{Y \vee Z}$ equiv $\overline{X \wedge Y} \wedge \overline{X \wedge Z}$.   (aa)

[Step 1: $\overline{X \wedge (Y \vee Z)}$ equiv $\overline{X} \vee \overline{Y \vee Z}$ by virtue of Equivalence 16 in accordance with Rule 1 (the variable Y is replaced with the formula $Y \vee Z$). Step 2: $\overline{(X \wedge Y) \vee (X \wedge Z)}$ equiv $\overline{X \wedge Y} \wedge \overline{X \wedge Z}$ by virtue of Equivalence 17 in accordance with Rule 1 (the variable X is replaced with the formula $X \wedge Y$ and the variable Y is replaced with the formula $X \wedge Z$). From a, Step 1, and Step 2, we obtain aa by virtue of properties of symmetry and transitivity of the equivalence relation.]

From aa and equivalences 17 and 16 we obtain, in accordance with Rules 1 and 2,

$$\bar{X} \vee (\bar{Y} \wedge \bar{Z}) \text{ equiv } (\bar{X} \vee \bar{Y}) \wedge (\bar{X} \vee \bar{Z}). \tag{aaa}$$

From aaa, we obtain, in accordance with Rule 1 (replacing $\bar{X}$ by X, $\bar{Y}$ by Y, and $\bar{Z}$ by Z),*

$X \vee (Y \wedge Z) \text{ equiv } (X \vee Y) \wedge (X \vee Z)$ (Equivalence 7).

(In what follows, we shall not indicate each time the rule according to which a formula is transformed but merely the equivalences used.)

In applications in which it is the function defined by a formula and not the structure of that formula itself that is of importance to us, we can replace one formula with another that is equivalent to it but that has a simpler structure. Thus, tranformation of formulas can be used, just as in ordinary algebra, to simplify them.

In long transformations of formulas, to obtain a more compact way of writing, we shall omit the symbol for conjunction and write, for example, simply XY instead of $X \wedge Y$.

Let us give an example of the simplification of a formula by means of transformations:

$(X\bar{Y} \vee Z)(\bar{X} \vee Y) \vee \bar{Z}$

equiv $(X\bar{Y} \vee Z)\bar{X} \vee (X\bar{Y} \vee Z)Y \vee \bar{Z}$	(6)
equiv $(X\bar{Y})\bar{X} \vee Z\bar{X} \vee (X\bar{Y})Y \vee ZY \vee \bar{Z}$	(6)
equiv $(X\bar{X})\bar{Y} \vee Z\bar{X} \vee X(\bar{Y}Y) \vee ZY \vee \bar{Z}$	(3, 5)
equiv $F\bar{Y} \vee Z\bar{X} \vee XF \vee ZY \vee \bar{Z}$	(15)
equiv $F \vee Z\bar{X} \vee F \vee ZY \vee \bar{Z}$	(12)
equiv $Z\bar{X} \vee ZY \vee \bar{Z}$	(10)
equiv $Z\bar{X} \vee Y \vee \bar{Z}$	(20)
equiv $Z\bar{X} \vee \bar{Z} \vee Y$	(2, 4)
equiv $\bar{X} \vee \bar{Z} \vee Y$.	(20)

* In actuality, we use Rule 1 to replace X, Y, Z by $\bar{X}$, $\bar{Y}$, $\bar{Z}$, respectively, and then we observe that $\bar{\bar{X}}$ is X, $\bar{\bar{Y}}$ is Y, and $\bar{\bar{Z}}$ is Z.

Thus, we have obtained

$(X\bar{Y} \vee Z)(\bar{X} \vee Y) \vee \bar{Z}$

equiv $\bar{X} \vee \bar{Z} \vee Y$.

These transformations are similar in many respects to the identities in ordinary algebra with disjunction, conjunction, T, and F corresponding respectively to addition, multiplication, the number 1, and the number 0. By virtue of the distributivity of conjunction with respect to disjunction, we can expand the expressions in parentheses, we can take a common factor outside the parentheses, we can omit summands equal to 0 and factors equal to 1, etc. In many other ways, the transformations are different; in particular, the transformations based on the distributivity of disjunction with respect to conjunction and on other properties of the operations of the algebra of propositions that the operations of ordinary algebra do not have.

EXERCISES

1.14. Simplify the following formulas:
a. $(X \vee Y) \wedge (X \vee \bar{Y})$
b. $X \vee (\bar{X} \wedge Y)$
c. $(X \wedge \bar{Y}) \vee (X \wedge \bar{Z}) \vee (Y \wedge Z) \vee Y \vee Z$
d. $(X \wedge Y \wedge Z) \vee (X \wedge Y \wedge \bar{Z}) \vee (X \wedge \bar{Y})$

For the equivalences that you obtain, write the dual equivalences.

1.15. Equivalences 16 and 17 can be written for the case of n variables as follows:

$$\overline{\bigwedge_{i=1}^{n} X_i} \text{ equiv } \bigvee_{i=1}^{n} \bar{X}_i \qquad (16')$$

$$\overline{\bigvee_{i=1}^{n} X_i} \text{ equiv } \bigwedge_{i=1}^{n} \bar{X}_i \qquad (17')$$

For $n = 2$, these equivalances have been proved (for example, with the aid of truth tables).

Prove them for arbitrary n by induction on the number of variables, that is, prove them for n under the assumption that they hold for $n-1$ variables (the method of mathematical induction in the algebra of logic).

1.16. Prove the equivalences
a. $A \lor (B \land C \land D)$ equiv $(A \lor B) \land (A \lor C) \land (A \lor D)$
b. $A \lor \bigwedge_{i=1}^{n} B_i$ equiv $\bigwedge_{i=1}^{n} (A \lor B_i)$

1.17. We have defined the five operations $\overline{}$, $\land$, $\lor$, $\Rightarrow$, $\Leftrightarrow$, each independently of the others. It turns out that we can confine ourselves to two operations and express the others in terms of these two. Find all pairs of operations that can be used for this purpose and show that the remaining three operations can be defined in terms of them. (Hint: each pair must include the operation $\overline{}$.)

1.18. Suppose that we have chosen as our basic operations disjunction, conjunction, and negation and that we define them not with the aid of the corresponding tables but with the aid of the dual equivalances 2 and 3, 6 and 7, 10 and 11, and 14 and 15. In other words, let us take these equivalences as our starting point.
  a. Prove equivalences 8 and 9. (Hint. Replace Y in 6 and 7 by X and replace Z by $\bar{X}$.)
  b. By using the original equivalences and equivalences 8 and 9 just proved, compile a truth table for disjunction, conjunction, and negation.

1.19. Suppose that we are given the following two compound propositions:
  a. If one of two terms is an integral multiple of 3 and if the sum of the two terms is an integral multiple of 3, then the other term is also an integral multiple of 3;
  b. If one of two terms is an integral multiple of 3 and the other is not an integral multiple of 3, then the sum is not an integral multiple of 3.

Replace the elementary propositions appearing in these compound propositions with variables and replace the words expressing the logical connectives by symbols for the corresponding operations. Show that the formulas obtained are equivalent.

Do the same for the propositions

c. If $a > b$ and ($b > 0$ or $b = 0$), then $a > 0$, and
d. if $a > b$ and $a \not> 0$, then $b \not> 0$ and $b \neq 0$.

1.20. Prove the equivalences

a. $X \Rightarrow (Y \Rightarrow Z)$ equiv $X \wedge Y \Rightarrow Z$
b. $(X \Rightarrow Y) \wedge (X \Rightarrow Z)$ equiv $X \Rightarrow Y \wedge Z$
c. $(X \Rightarrow Z) \vee (Y \Rightarrow Z)$ equiv $X \wedge Y \Rightarrow Z$

Give specific examples of the propositions expressed with the aid of these formulas.

2.4. From the standpoint of the values assumed by these formulas the set of all possible formulas of the algebra of propositions can be divided into three classes.

1. Formulas assuming the value T for all combinations of values of the variables appearing in them are said to be **identically true formulas** or **tautologies**.

2. Formulas that assume the value F for all combinations of values of the variables appearing in them are called **identically false formulas**. (Their negations are tautologies.)

3. Finally, formulas that assume the value T for at least one combination of the variables appearing in them and the value F for at least one combination are often called **contingent** formulas.

A special role in the algebra of propositions is played by tautologies, which express the *laws of logic* in the language of that algebra.

The proposition "φ is a tautology" is usually indicated by $\vDash \varphi$. (Of course, the symbol $\vDash$ and the notation for the (meta) proposition $\vDash \varphi$ do not belong to the propositional calculus itself but to its metalanguage.*)

If two formulas in the algebra of propositions φ_1 and φ_2 are equivalent, that is, if φ_1 equiv φ_2, then, by combining these formulas with the symbol $\Leftrightarrow$, we obtain a tautology in the algebra of propositions $\varphi_1 \Leftrightarrow \varphi_2$. Conversely, if $\vDash \varphi_1 \Leftrightarrow \varphi_2$, then φ_1 equiv φ_2.

To see this, suppose that φ_1 equiv φ_2. Then, φ_1 and φ_2 assume

* See footnote, p. 48.

the same values for any combination of the values of the variables appearing in them, and, in such a case, the formula $\varphi_1 \Leftrightarrow \varphi_2$, in accordance with the definition of the operation $\Leftrightarrow$, assumes only the value T, so that this formula is a tautology.

For the converse, suppose that $\vDash \varphi_1 \Leftrightarrow \varphi_2$. Then, for any combination of values of the variables, both φ_1 and φ_2 assume the value T or both assume the value F; that is, formulas φ_1 and φ_2 assume the same values, so that φ_1 equiv φ_2.

Thus, each of the relationships (1–21) enumerated previously "generates" a tautology in the algebra of propositions expressing some law of logic.

Below is a list of the more important tautologies in the algebra of propositions, formulas that find extensive application.

We reserve the first nineteen numbers for tautologies generated by relations 1–21:

1. $\bar{\bar{X}} \Leftrightarrow X$ — law of double negation

2. $X \vee Y \Leftrightarrow Y \vee X$ — law of commutativity of disjunction

3. $X \wedge Y \Leftrightarrow Y \wedge X$ — law of commutativity of conjunction

4. $(X \vee Y) \vee Z \Leftrightarrow X \vee (Y \vee Z)$ — law of associativity of disjunction

5. $(X \wedge Y) \wedge Z \Leftrightarrow X \wedge (Y \wedge Z)$ — law of associativity of conjunction

6. $X \wedge (Y \vee Z) \Leftrightarrow (X \wedge Y) \vee (X \wedge Z)$ — law of distributivity of conjunction with respect to disjunction

7. $X \vee (Y \wedge Z) \Leftrightarrow (X \vee Y) \wedge (X \vee Z)$ law of distributivity of disjunction with respect to conjunction

8. $X \vee X \Leftrightarrow X$ ⎫
9. $X \wedge X \Leftrightarrow X$ ⎬   idempotency laws

10. $X \vee F \Leftrightarrow X$

11. $X \wedge T \Leftrightarrow X$

12.* $\overline{X \wedge F}$

13. $X \vee T$

14. $X \vee \overline{X}$ law of the excluded middle

15. $\overline{X \wedge \overline{X}}$ law of contradiction

16. $\overline{X \wedge Y} \Leftrightarrow \overline{X} \vee \overline{Y}$ ⎫
17. $\overline{X \vee Y} \Leftrightarrow \overline{X} \wedge \overline{Y}$ ⎬   de Morgan's laws

18. $(X \Rightarrow Y) \Leftrightarrow \overline{X} \vee Y$

19. $(X \Leftrightarrow Y) \Leftrightarrow (X \Rightarrow Y) \wedge (Y \Rightarrow X)$

In addition to these laws, with which we are essentially familiar, we give a few others:

20. $X \Leftrightarrow X$ identity law

21. $(X \Rightarrow Y) \wedge (Z \Rightarrow Y) \Leftrightarrow (X \vee Z \Rightarrow Y)$

22. $(X \Rightarrow Y) \wedge (X \Rightarrow Z) \Leftrightarrow (X \Rightarrow Y \wedge Z)$

23. $(X \Rightarrow Y) \Leftrightarrow (\overline{Y} \Rightarrow \overline{X})$ contraposition law

* In formulas 12–15, we omitted the symbol $\Leftrightarrow$. Since, for example, $X \wedge F$ equiv F, it follows that $\overline{X \wedge F}$ equiv $\overline{F}$ that is, $\overline{X \wedge F}$ equiv T. The formula $\overline{X \wedge F}$ assumes only the value T. Therefore, $\vDash \overline{X \wedge F}$. Analogously, instead of writing $X \vee \overline{X} \Leftrightarrow T$, we write simply $X \vee \overline{X}$.

24. $(X \wedge Y \Rightarrow Z) \Leftrightarrow (X \wedge \bar{Z} \Rightarrow \bar{Y})$ extended contraposition law

25. $[X \Rightarrow (Y \Rightarrow Z)] \Leftrightarrow (X \wedge Y \Rightarrow Z)$

26. $(X \Rightarrow Y) \wedge X \Rightarrow Y$

27. $(X \Rightarrow Y) \wedge \bar{Y} \Rightarrow \bar{X}$

28. $(X \vee Y) \wedge \bar{X} \Rightarrow Y$

29. $(X \Rightarrow Y) \wedge (Y \Rightarrow Z) \Rightarrow (X \Rightarrow Z)$

30. $(X \Rightarrow Y \wedge \bar{Y}) \Rightarrow \bar{X}$. syllogism law

2.5. In view of the special role of tautologies expressing the laws of logic, it is important for us to be able to tell whether an arbitrary specific formula in the algebra of propositions is or is not a tautology.

The problem arises of finding a general method enabling us to answer the question (posed for a specific formula in the algebra of propositions), "Is this formula a tautology?"

A method that allows us to answer yes or no for an arbitrary case of a general question is called a **decision method** or an **algorithm** or a **decision procedure** for this question. The problem of finding such a method is called the **decision problem** for the question.

For the question posed above, this problem has an affirmative answer. The decision procedure can be the construction of a truth table which, for any given formula, always enables us to answer yes or no to the question. If the right column of the table, corresponding to the given formula, contains only T, the formula is a tautology. On the other hand, if at least one F appears, it is not a tautology.

Of course, this method is not always practical because, when there are a large number of variables, compiling the truth table is very laborious. However, it always consists of a finite number of steps and, in theory, always provides the answer to the question.

In Section 4, we shall present another decision procedure, based on the reduction of formulas to a standard or normal form.

EXERCISES

1.21. Prove by suitable transformations that formulas 21–30 are tautologies.

Suggestions: In 21–25, the formula in question has the form $\varphi_1 \Leftrightarrow \varphi_2$. One need only show that φ_1 equiv φ_2. In 26–30, the formula has the form of an implication. Transform it to an equivalent formula containing only the symbols for negation, conjunction, and disjunction. Do this in such a way that the negation symbols apply to the elementary propositions.

1.22. With the aid of truth tables and transformations, prove that the following formulas from the algebra of propositions are tautologies:
a. $X \Rightarrow (Y \Rightarrow X)$;
b. $[X \Rightarrow (Y \Rightarrow Z)] \Rightarrow [(X \Rightarrow Y) \Rightarrow (X \Rightarrow Z)]$;
c. $X \wedge Y \Rightarrow X$;
d. $(X \Rightarrow Y) \Rightarrow [(X \Rightarrow Z) \Rightarrow (X \Rightarrow (Y \wedge Z))]$;
e. $X \Rightarrow X \vee Y$;
f. $(X \Rightarrow Z) \Rightarrow [(Y \Rightarrow Z) \Rightarrow (X \vee Y \Rightarrow Z)]$.

1.23. Let X/Y denote the result of the operation (known as Sheffer's operation) on two propositions X and Y, defined as follows: The proposition X/Y is false if and only if X and Y are both true. Set up a truth table defining X/Y depending on the values of the variables X and Y. Prove the following equivalences:
a. X/Y equiv $\bar{X} \vee \bar{Y}$;
b. X/Y equiv $X \Rightarrow \bar{Y}$;
c. X/Y equiv Y/X;
d. $\bar{X}$ equiv X/X;
e. $\overline{X/Y}$ equiv $X \wedge Y$.

3. EXAMPLES OF THE APPLICATION OF THE LAWS OF THE LOGIC OF PROPOSITIONS IN DERIVATIONS

In our derivation of certain propositions from others, we use the laws of logic.

In Section 2, we pointed out that the tautologies in the algebra of propositions express laws of logic. Of course, the laws of logic expressed by means of the algebra of propositions do not exhaust all the laws of logic that are used in our reasoning. In particular, if in the derivation of certain propositions from others we consider not only the structure of the compound propositions but also the internal logical structure of the elementary propositions comprising them, we use laws of logic that cannot be expressed by means of the algebra of propositions, which does not take into account the internal structure of the elementary propositions. Thus, the laws of logic expressed by the tautologies of the algebra of propositions can serve as a basis only for those derivations in which we consider only the structure of compound propositions and not the structure of the propositions considered in propositional logic as elementary.

When we disregard the content of such reasoning, that is, when we replace the elementary propositions in our reasoning with variables, we obtain the following outline of a derivation:

From $\varphi_1, \varphi_2, \ldots, \varphi_n$ it follows that (we derive) φ.*

This should be understood as follows: "If the propositions with structure expressed by formulas $\varphi_1, \varphi_2, \ldots, \varphi_n$ are true (the premises), then the proposition with structure expressed by the formula φ is true (the conclusion)." The most significant point here is that in the derivation we consider only the structure of the premises and conclusion and disregard their content.

The scheme, or rule, of derivation (**inference** in the terminology of traditional logic) with premises $\varphi_1, \varphi_2, \ldots, \varphi_n$ and conclusion φ is written as follows:

$$\frac{\varphi_1, \varphi_2, \ldots, \varphi_n}{\varphi}.$$

This rule of inference is admissible and the reasoning in which it is used is correct if the implication $\bigwedge_{i=1}^{n} \varphi_i \Rightarrow \varphi$ is a tautology in the

---

* For brevity, we do not indicate here the variables appearing in the composition of the formulas; that is, instead of $\varphi(X_1, X_2, \ldots, X_n)$, we write simply φ.

algebra of propositions ($\vDash \bigwedge_{i=1}^{n} \varphi_i \Rightarrow \varphi$), that is, if it expresses a law of propositional logic. If $\vDash \bigwedge_{i=1}^{n} \varphi_i \Rightarrow \varphi$, then, under the assumption that the premises are true, and hence that their conjunction ($\bigwedge_{i=1}^{n} \varphi_i$) is true, the conclusion φ will also be a true proposition, because otherwise we would have a combination of values of the variables in $\varphi_1, \varphi_2, \ldots, \varphi_n, \varphi$ such that the implication $\bigwedge_{i=1}^{n} \varphi_i \Rightarrow \varphi$ would assume the value F, that is, would not be a tautology.

Let us give some examples of reasoning in which one uses rules of inference based on some of the laws of propositional logic listed in Section 2.

3.1. Let us make a logical analysis of the following proposition: "If a given polygon is regular, then it is possible to inscribe a circle in it. A certain polygon is regular; consequently, it is possible to inscribe a circle in that polygon."

In this reasoning, a certain rule of inference is used. The word *consequently* is usually used to separate the premises from the conclusion. To make clear the crux of the rule used in this reasoning, we ignore the content of the elementary propositions that appear in it. We replace the elementary proposition "the given polygon is regular" with the variable X and we replace the proposition "it is possible to inscribe a circle in it (i.e., in the given polygon)" with the variable Y. Then, the scheme of the reasoning that we are studying is written in the following form:

$$\frac{X \Rightarrow Y, X}{Y};$$

that is, from the propositions $X \Rightarrow Y$ and X, we derive the conclusion Y.

This rule of inference is admissible: the implication $(X \Rightarrow Y) \wedge X \Rightarrow Y$, which is one of the tautologies of the algebra of propositions, expresses a law of propositional logic (Formula 26 on p. 61).

In traditional logic, the inference $\dfrac{X \Rightarrow Y,\ X}{Y}$ is called *modus ponens*. From the assertion that X is true together with the premise $X \Rightarrow Y$ we proceed to the assertion that Y is true.

In contemporary logic, this rule of inference is also called the rule of detachment (because from the premise $X \Rightarrow Y$ we detach the conclusion Y with the aid of the premise X), and, in many logical calculi (axiomatically constructed logical systems), it is taken as the basic rule of inference which is used with a system of axioms.

3.2. Let us now look at the following reasoning: "If a given polygon is regular, it is possible to inscribe a circle in it. A certain polygon is such that one cannot inscribe a circle in it. Consequently, the given polygon is not regular."

Using the symbolic notation introduced above, we obtain the following scheme for this reasoning:

$$\dfrac{X \Rightarrow Y,\ \overline{Y}}{\overline{X}}.$$

The admissability of this rule of inference follows from Formula 27.

In traditional logic, the inference $\dfrac{X \Rightarrow Y,\ \overline{Y}}{\overline{X}}$ is called *modus tollens*. From denial of the truth of Y we arrive with the aid of the premise $X \Rightarrow Y$ at the denial of the truth of X.

Remark: The rules of inference given in Sections 3.1 and 3.2 enable us in the case of a true implication to conclude from the truth of the premise that the consequence is true and from the falsity of the consequence that the premise is false.

In connection with the definition of implication, we have also shown that from the falsity of the premise we cannot conclude that the conclusion is false nor can we conclude from the truth of the conclusion that the premise is true. Thus, both of the following lines of reasoning are incorrect:

If a given polygon is regular, it is possible to inscribe a

circle in it. A certain polygon is not regular. Consequently, it is not possible to inscribe a circle in it.

and

If a given polygon is regular, it is possible to inscribe a circle in it. A certain polygon is such that one can inscribe a circle in it. Consequently, this polygon is regular.

In these two incorrect lines of reasoning, the conclusions are drawn in accordance with the following schemes:

$$\frac{X \Rightarrow Y, \bar{X}}{\bar{Y}} \quad \text{and} \quad \frac{X \Rightarrow Y, Y}{X}$$

It is easy to show that the implications

$$(X \Rightarrow Y) \wedge \bar{X} \Rightarrow \bar{Y} \quad \text{and} \quad (X \Rightarrow Y) \wedge Y \Rightarrow X$$

are not tautologies in the algebra of propositions:

X	Y	$X \Rightarrow Y$	$\bar{X}$	$(X \Rightarrow Y) \wedge \bar{X}$	$\bar{Y}$	$(X \Rightarrow Y) \wedge \bar{X} \Rightarrow \bar{Y}$
T	T	T	F	F	F	T
F	T	T	T	T	F	F

We do not need to continue this table further. It is sufficient to find a single set of values of the variables for which the implication in question is false in order to draw the conclusion that it is not a tautology.

3.3. Let us look at the following reasoning:

If a number is rational, it can be represented as the ratio of two integers; consequently, if a number cannot be represented as the ratio of two integers, it is not rational.

Let us replace the elementary propositions in this reasoning by variables. Let us replace the proposition "the number is rational" by the variable X and the proposition "the number can be represented as a ratio of two integers" by Y.

Then, the rule of inference that is used in the above reasoning can be written

$$\frac{X \Rightarrow Y}{\overline{Y} \Rightarrow \overline{X}}.$$

This rule, called the rule of contraposition, is based on the law of contraposition (23), $X \Rightarrow Y \Leftrightarrow \overline{Y} \Rightarrow \overline{X}$, or actually on part of it, $(X \Rightarrow Y) \Rightarrow (\overline{Y} \Rightarrow \overline{X})$. Since
$X \Rightarrow Y \Leftrightarrow \overline{Y} \Rightarrow \overline{X}$ equiv $((X \Rightarrow Y) \Rightarrow (\overline{Y} \Rightarrow \overline{X}))$
$\wedge ((\overline{Y} \Rightarrow \overline{X}) \Rightarrow (X \Rightarrow Y))$,
the law of contraposition can be represented as the two laws

$(X \Rightarrow Y) \Rightarrow (\overline{Y} \Rightarrow \overline{X})$ and $(\overline{Y} \Rightarrow \overline{X}) \Rightarrow (X \Rightarrow Y)$.

The rule of contraposition is frequently applied in mathematics in connection with the study of a system of interrelated theorems (direct, converse, negative, and negative converse):

If $X \Rightarrow Y$ expresses the direct theorem,

then $Y \Rightarrow X$ expresses the converse,

$\overline{X} \Rightarrow \overline{Y}$ expresses the negative,

and $\overline{Y} \Rightarrow \overline{X}$ expresses the negative converse.

On the basis of the law of contraposition (23), if we have proven the direct theorem, then the negative converse of that theorem, which is equivalent to it, will also be true. Analogously, from the converse theorem we derive the negative and vice versa.

If we apply the rule of contraposition to $\overline{X} \Rightarrow \overline{Y}$:

$$\frac{\overline{X} \Rightarrow \overline{Y}}{\overline{\overline{Y}} \Rightarrow \overline{\overline{X}}}$$

we obtain the conclusion $\overline{\overline{Y}} \Rightarrow \overline{\overline{X}}$.

If we now apply to $\overline{\overline{Y}} \Rightarrow \overline{\overline{X}}$ the law of double negation (1), $\overline{\overline{X}} \Leftrightarrow X$, that is, if we replace $\overline{\overline{Y}}$ and $\overline{\overline{X}}$ with the equivalent formulas Y and X, we obtain $Y \Rightarrow X$.

3.4. Let us look at the following reasoning: "We know that if a number is an integral multiple of both 2 and 3, it is an integral multiple of 6. Consequently, if a number is an integral multiple of 2 and is not an integral multiple of 6, it is not an integral multiple of 3."

What rule of inference is used in this reasoning? Let us ignore the specific meaning of the elementary propositions here and replace the proposition "a number is an integral multiple of 2" by the letter X, the proposition "a number is an integral multiple of 3" by Y, and "a number is an integral multiple of 6" by Z. Then, the premise may be written in the form

$$X \wedge Y \Rightarrow Z$$

and the conclusion in the form

$$X \wedge \bar{Z} \Rightarrow \bar{Y}.$$

Thus, verification of the validity of the reasoning in question or the admissibility of the rule of inference

$$\frac{X \wedge Y \Rightarrow Z}{X \wedge \bar{Z} \Rightarrow \bar{Y}} \quad \left(\text{or } \frac{X \wedge Y \Rightarrow Z}{\bar{Z} \wedge Y \Rightarrow \bar{X}}\right)$$

reduces to proving that the implication $(X \wedge Y \Rightarrow Z) \Rightarrow (X \wedge \bar{Z} \Rightarrow \bar{Y})$, or, correspondingly, the implication $(X \wedge Y \Rightarrow Z) \Rightarrow (\bar{Z} \wedge Y \Rightarrow \bar{X})$, is a tautology.

The fact that this implication is a tautology follows from the law of extended contraposition:

$$(X \wedge Y \Rightarrow Z) \Leftrightarrow (X \wedge \bar{Z} \Rightarrow \bar{Y}).$$

This rule of inference is also called the rule of extended contraposition. In its most general form, this rule reads

$$\frac{X_1 \wedge X_2 \wedge \ldots \wedge X_k \wedge \ldots \wedge X_n \Rightarrow Z}{X_1 \wedge X_2 \wedge \ldots \wedge \bar{Z} \wedge \ldots \wedge X_n \Rightarrow \bar{X}_k}$$

and the corresponding law is

$$(X_1 \wedge X_2 \wedge \ldots \wedge X_k \wedge \ldots \wedge X_n \Rightarrow Z)$$
$$\Leftrightarrow (X_1 \wedge X_2 \wedge \ldots \wedge \bar{Z} \wedge \ldots \wedge X_n \Rightarrow \bar{X}_k).$$

3.5. Let us make a logical analysis of the following line of reasoning: "If a triangle is isosceles, two of its sides are equal. If two sides of a triangle are equal, two of its angles are equal.

Consequently, if a triangle is isosceles, two of its angles are equal."

Here, we draw a conclusion from two premises. To answer the question as to the admissibility of the rule of inference used in this reasoning, we replace the specific elementary propositions by variables. We denote the proposition "a triangle is isosceles" by the letter X, the proposition "two sides of a triangle are equal" by the letter Y, and the proposition "two angles are equal" by the letter Z.

Then, the first premise is written in the form of the implication $X \Rightarrow Y$, the second in the form $Y \Rightarrow Z$, and the conclusion in the form $X \Rightarrow Z$. Here, the rule of inference that we are using is

$$\frac{X \Rightarrow Y,\ Y \Rightarrow Z}{X \Rightarrow Z}.$$

This rule of inference, known as the syllogism rule, is extensively used in mathematical proofs, and is based, as one can easily see, on the law of propositional logic

$$(X \Rightarrow Y) \wedge (Y \Rightarrow Z) \Rightarrow (X \Rightarrow Z), \qquad \text{(Formula 2a)}$$

which is known by the same name.

3.6 By analyzing lines of reasoning by means of the algebra of propositions, one can easily point out the error of certain deductions.

Let us look at an example: "In a parallelogram, the opposite sides are pairwise parallel; in a rhombus, the opposite sides are also pairwise parallel. Consequently, a rhombus is a parallelogram.'

The conclusion "a rhombus is a parallelogram" is a true proposition. However, neither the true proposition "a rhombus is a parallelogram" nor the false proposition "a parallelogram is a rhombus" (meaning here that "every parallelogram is a rhombus") follows from the given premises and the reasoning is erroneous. (The incorrectness here is known in traditional logic by the name *non sequitur*, meaning "it does not follow.") The error in this reasoning is easily spotted if we express it in the language of the algebra of propositions.

Let us, as a preliminary, connect the premises and the conclusion with the words "if..., then":

"If a quadrilateral is a parallelogram, then its opposite sides are pairwise parallel. If a quadrilateral is a rhombus, then its opposite sides are pairwise parallel. Consequently, if a quadrilateral is a rhombus, then it is a parallelogram."

Let us denote the proposition "a quadrilateral is a parallelogram" by the letter X, the proposition "a quadrilateral is a rhombus" by the letter Y, and the proposition "the opposite sides of a quadrilateral are pairwise parallel" by the letter Z.

Now, we need to verify whether the conclusion $Y \Rightarrow X$ (or the conclusion $X \Rightarrow Y$) follows from the premises $X \Rightarrow Z$ and $Y \Rightarrow Z$, that is, whether the rule of inference

$$\frac{X \Rightarrow Z,\ Y \Rightarrow Z}{Y \Rightarrow X}$$

is valid.

Such a rule of inference is admissible in our reasoning if the implication

$$(X \Rightarrow Z) \wedge (Y \Rightarrow Z) \Rightarrow (Y \Rightarrow X) \tag{a}$$

is a tautology in the algebra of propositions.

However, this formula is not a tautology. To show this, we need only find one set of values for the variables X, Y, and Z for which it assumes the value F, that is, for which it becomes a false proposition. Such a set of values is (F, T, T). When we substitute these values for X, Y, and Z in implication (a), we obtain

$(F \Rightarrow T) \wedge (T \Rightarrow T) \Rightarrow (T \Rightarrow F)$ equiv $T \wedge T \Rightarrow F$;
equiv $T \Rightarrow F$;
equiv F.

EXERCISES

1.24. Carry out a logical analysis of the following lines of reasoning; that is, replace the elementary propositions in them by variables and show the admissibility or inadmissibility of the rules of inference used in them:

1. If a number ends in 0, it is divisible by 5. A certain number ends in 0. Consequently, it is divisible by 5.

2. If a number ends in 0, it is divisible by 5. A certain number does not end in 0. Consequently, it is not divisible by 5.

3. If a number ends in 0, it is divisible by 5. A certain number is divisible by 5. Consequently, it ends in 0.

4. If a number ends in 0, it is divisible by 5. A certain number is not divisible by 5. Consequently, it does not end in 0.

5. If the diagonals of a parallelogram are perpendicular to each other, that parallelogram is a rhombus. The diagonals of a certain parallelogram are not perpendicular to each other. Consequently, the given parallelogram is not a rhombus.

6. Every fraction is a rational number. Every integer is a rational number. Consequently, every integer is a fraction.

7. Every fraction is a rational number. Every integer is a rational number. Consequently every fraction is an integer.

8. If a number is an integer, it is rational. If a number is an irreducible fraction, it is not an integer. Consequently, if a number is an irreducible fraction, it is not a rational number.

9. If the premises in a certain line of reasoning are true and if the rule of inference used is admissible, then the conclusion is true. In a certain line of reasoning, the conclusion is false. Consequently, either the premises are false or an admissible rule of inference was not used in the reasoning.

10. If $a = 0$ or $b = 0$, then $ab = 0$. But $ab \neq 0$; consequently, $a \neq 0$ and $b \neq 0$.

1.25. Write the rules of inference based on laws 21, 22, 25, 28, and 30 of Section 2.7, and give specific examples of reasoning in which these rules of inference are used.

1.26. Check the admissibility of deductions in accordance with the following schemes:

a. $\dfrac{X \vee Y, \bar{X}}{Y}$

b. $\dfrac{X \vee Y, X}{\bar{Y}}$

c. $\dfrac{X \dot{\vee} Y, X}{\bar{Y}}$

d. $\dfrac{X \vee Y \vee Z, \bar{X} \wedge \bar{Y}}{Z}$

e. $\dfrac{X \vee Y \vee Z, X}{\bar{Y} \wedge \bar{Z}}$

f. $\dfrac{X \vee Y \vee Z, \bar{X}}{Y \vee Z}$

Give examples of reasoning in which deductions in accordance with these schemes are made.

1.27. Prove

a. $\vDash ([(X \Rightarrow Y) \Rightarrow (Z \wedge W)] \wedge \bar{Z}) \Rightarrow (X \wedge \bar{Y})$
b. $\vDash [(X \Rightarrow Y) \wedge (Z \Rightarrow W) \wedge (X \vee Z)] \Rightarrow Y \vee W$
c. $\vDash [(X \Rightarrow Y) \wedge (Z \Rightarrow W) \wedge (\bar{Y} \wedge \bar{W})] \Rightarrow \bar{X} \wedge \bar{Z}$

Write rules of inference based on these laws and give examples of reasoning in which deductions are made in accordance with them.

4. NORMAL FORMS OF FUNCTIONS. MINIMAL FORMS

In this section, we shall assume that the variables A, B, C, ..., X, Y, and Z in the algebra of propositions can assume the two values 1 and 0, without interpreting these values as truth values or giving them any other specific interpretation.

In the two-element set $\{0, 1\}$, we define operations which we shall call "negation," "conjunction," and "disjunction" just as in a meaningful* algebra of propositions, but here we shall not assign to them any specific meaning and shall proceed only on the basis of their formal definitions, which are as follows:

1. $\bar{X} = \begin{cases} 0 \text{ if } X = 1, \\ 1 \text{ if } X = 0. \end{cases}$

2. $X \wedge Y = \begin{cases} 1 \text{ if } X = 1 \text{ and } Y = 1, \\ 0 \text{ otherwise.} \end{cases}$

---

* We use the word "meaningful" to indicate that the meaning or content of symbols and expressions is taken into consideration, as opposed to the study of their formal properties. The technical term "contensive" might be more appropriate than "meaningful".

3. $X \vee Y = \begin{cases} 0 \text{ if } X = 0 \text{ and } Y = 0, \\ 1 \text{ otherwise.} \end{cases}$

The conjunction $X \wedge Y$ is also called the **product**, abbreviated as XY, and X and Y are called **factors**. The disjunction $X \vee Y$ is also called the **sum**, and X and Y are called **summands**. (Sometimes, we use the usual symbol $X + Y$ or a modification of it, $X \oplus Y$, for the sum.)

One can easily see that if we replace 1 with T and 0 with F, these definitions coincide with the definitions of these same operations in the "meaningful" algebra of propositions (see Section 1). From this it follows that all the equivalences of formulas 1–17 (see Section 2.3) that were established with the aid of the symbols for the three operations, $\bar{}$, $\wedge$, $\vee$ remain valid in our new "abstract" algebra.* (Here, we denote the equivalence of formulas with the usual equal sign $=$.)

The functions in this abstract algebra (we shall call it **Boolean algebra** and the functions in it **Boolean functions**) are not to be interpreted as propositional functions (propositions mapped into true or false propositions) but simply as binary functions of binary arguments, defined on the set $\{0, 1\}^n$ (functions of n variables) and assuming values in $\{0, 1\}$.

The "meaningful" algebra of propositions expounded in Sections 1 and 2 is one of the possible models of abstract Boolean algebra.

In this section, we shall use basically the terminology and symbols of abstract Boolean algebra, but occasionally we shall also use expressions relating to the language of the meaningful algebra of propositions, for example, "the conjunction is true" instead of "the conjunction has value 1." (In mathematics, just as in any abstract theory, we frequently use expressions relating to the language of some specific model.)

---

* Other operations in the algebra of propositions can be reduced to these three with the aid of equivalences 18 and 19.
† We assume that the reader is familiar, at least in its broad outlines, with the concept of a model (in its application to axiomatic systems, for example, to geometry). For greater detail, see Church [1], Section 07–Section 09.

4.1. From the table defining a function, one can construct a formula expressing that function.

Let us look at a specific example. Suppose that a function f of three variables is defined by the following table:*

X	Y	Z	$f(X, Y, Z)$
0	0	0	0
0	0	1	0
0	1	0	0
0	1	1	1
1	0	0	0
1	0	1	1
1	1	0	1
1	1	1	1

We shall indicate two ways of constructing a formula determining this function.

Method 1: a. Let us choose the combinations of values of the variables for which $f(X, Y, Z) = 1$:

(0, 1, 1); (1, 0, 1); (1, 1, 0); (1, 1, 1).

b. To each of these combinations, we assign the conjunction of the variables X, Y, Z or of their negations that assumes for those values of the variables the value 1. Thus,

the triple (0, 1, 1) corresponds to $\bar{X}YZ$,
the triple (1, 0, 1) corresponds to $X\bar{Y}Z$,
the triple (1, 1, 0) corresponds to $XY\bar{Z}$,
the triple (1, 1, 1) corresponds to XYZ.

c. The disjunction of these conjunctions is equal to 1 if and only if the given function assumes the value 1 and hence represents one of the possible ways of expressing that function:

* In this and in the following tables, the combinations of values of the variables are arranged in such a way that if we regard them as binary numbers (in the present case, three-digit), these numbers are arranged in increasing size (from zero to seven).

$f(X, Y, Z) = \bar{X}YZ \vee X\bar{Y}Z \vee XY\bar{Z} \vee XYZ.$

To see this, let us take an arbitrary triple of the variables $(\alpha_1, \alpha_2, \alpha_3)$, where $\alpha_i = 0$ or 1. If, in accordance with the table, $f(\alpha_1, \alpha_2, \alpha_3) = 1$, then in the formula that we obtain one of the terms of the disjunction will assume the value 1 for this combination of values of the variables; that is, the entire disjunction assumes the value 1.

If in accordance with the table $f(\alpha_1, \alpha_2, \alpha_3) = 0$, then, in the disjunction obtained, all the terms assume the value 0 (since each of them assumes the value 1 only when f assumes the value 1). Consequently, the disjunction itself assumes the value 0.

Thus, the formula obtained does indeed define the same function as the table above; that is, it is one of the possible ways of defining that function. This expression is called the **perfect disjunctive normal form** of the given function.

Method 2: a. Let us choose combinations of the values of the variables for which $f(X, Y, Z) = 0$:

(0, 0, 0); (0, 0, 1); (0, 1, 0); (1, 0, 0).

b. To each of these combinations, let us assign the disjunction of the variables X, Y, Z, or their negations that assumes for these values of the variables the value 0. Thus,

the triple (0, 0, 0) corresponds to $X \vee Y \vee Z$,
the triple (0, 0, 1) corresponds to $X \vee Y \vee \bar{Z}$,
the triple (0, 1, 0) corresponds to $X \vee \bar{Y} \vee Z$,
the triple (1, 0, 0) corresponds to $\bar{X} \vee Y \vee Z$.

c. The conjunction of these disjunctions is equal to 0 if and only if the given function assumes the value 0, and consequently is one of the possible ways of expressing this function:

$$f(X, Y, Z) = (X \vee Y \vee Z) \wedge (X \vee Y \vee \bar{Z}) \wedge (X \vee \bar{Y} \vee Z)$$
$$\wedge (\bar{X} \vee Y \vee Z).$$

(We leave the proof of this to the reader.) This expression (the dual of the perfect disjunctive normal form) is called the **perfect conjunctive normal form** of the given function.

To have a more compact notation for these two normal forms, we introduce the following notation:

$$X^\alpha = \begin{cases} X & \text{if } \alpha = 1, \\ \bar{X} & \text{if } \alpha = 0. \end{cases}$$

Then, the conjunction corresponding to an arbitrary triple $(\alpha_1, \alpha_2, \alpha_3)$, where $\alpha_i = 0$ or 1, of the values of the variables X, Y, Z (that is, the conjunction which is true, or assumes the value 1, for these values of the variables) is written $X^{\alpha_1} Y^{\alpha_2} Z^{\alpha_3}$. The corresponding disjunction (that is, the disjunction which is false or assumes the value 0 for these values of the variables) is written $X^{\bar{\alpha}_1} \vee Y^{\bar{\alpha}_2} \vee Z^{\bar{\alpha}_3}$ where

$$\bar{\alpha}_i = \begin{cases} 0 & \text{if } \alpha_i = 1, \\ 1 & \text{if } \alpha_i = 0. \end{cases}$$

In this notation, the perfect disjunctive normal form is written as follows:

$$f(X, Y, Z) = \bigvee_1 X^{\alpha_1} Y^{\alpha_2} Z^{\alpha_3},$$

where $\bigvee_1$ denotes disjunction over all those triples $(\alpha_1, \alpha_2, \alpha_3)$ of values of the variables for which the function is equal to 1, and the perfect conjunctive normal form is written

$$f(X, Y, Z) = \bigwedge_0 (X^{\bar{\alpha}_1} \vee Y^{\bar{\alpha}_2} \vee Z^{\bar{\alpha}_3}),$$

where $\bigwedge_0$ denotes conjunction of all those triples $(\alpha_1, \alpha_2, \alpha_3)$ of values of the variables for which the function is equal to 0.

A conjunction of the form $X_1^{\alpha_1} X_2^{\alpha_2} \ldots X_n^{\alpha_n}$, the factors of which are variables or their negations, is called an **elementary conjunction** of rank n.

A disjunction of the form $X_1^{\alpha_1} \vee X_2^{\alpha_2} \vee \ldots \vee X_n^{\alpha_n}$, the terms of which are variables or their negations, is called an **elementary disjunction** of rank n.

Thus, the perfect disjunctive normal form of a function of n variables is a disjunction whose terms are elementary conjunctions of rank n, and a perfect conjunctive normal form is a conjunction the factors of which are elementary disjunctions of rank n.

Let us show that an arbitrary function of n variables that is not identically equal to 0 can be represented as a perfect disjunctive normal form.

First, we shall show that a function $f(X_1, X_2, \ldots, X_n)$ can be represented in the following form:

$$f(X_1, X_2, \ldots, X_n) = \bigvee_{(\alpha_1, \ldots, \alpha_n)} X_1^{\alpha_1} X_2^{\alpha_2} \ldots X_n^{\alpha_n} f(\alpha_1, \alpha_2, \ldots, \alpha_n). \quad (1)$$

(The symbol $\bigvee_{(\alpha_n, \ldots, \alpha_1)}$ should be understood to mean disjunction over all possible n-tuples $(\alpha_1, \alpha_2, \ldots, \alpha_n)$, where $\alpha_i = 0$ or 1.)

As a preliminary, we note that

$$X_i^{\alpha_i} = \begin{cases} 1 \text{ if } X_i = \alpha_i, \\ 0 \text{ if } X_i \neq \alpha_i. \end{cases}$$

This is true because, if $X_i = \alpha_i = 0$, then $X_i^{\alpha_i} = \bar{X}_i = \bar{0} = 1$; if $X_i = \alpha_i = 1$, then $X_i^{\alpha_i} = X_i = 1$; if $X_i = 1$ and $\alpha_i = 0$, then $X_i^{\alpha_i} = \bar{X}_i = \bar{1} = 0$; and if $X_i = 0$ and $\alpha_i = 1$, then $X_i^{\alpha_i} = X_i = 0$.

Thus, the elementary conjunction $X_1^{\alpha_1} X_2^{\alpha_2} \ldots X_n^{\alpha_n}$ fails to be equal to 0 if and only if the n equations $X_i = \alpha_i$, for $i = 1, 2, \ldots, n$, hold simultaneously, that is, if and only if $X_1 = \alpha_1$, $X_2 = \alpha_2$, $\ldots$, $X_n = \alpha_n$.

To prove Equation 1, we take an arbitrary n-tuple $(\alpha_1', \alpha_2', \ldots, \alpha_n')$ of values of the arguments $X_1, X_2, \ldots, X_n$. The left-hand member of the equation becomes $f(\alpha_1', \alpha_2', \ldots, \alpha_n')$. On the right-hand side, all the terms of the disjunction in which $\alpha_i' \neq \alpha_i$ for at least one i vanish. There remains only the term of the disjunction in which $\alpha_i' = \alpha_i$ for all i and $X_1^{\alpha_1} X_2^{\alpha_2} \ldots X_n^{\alpha_n}$ takes the value 1. Consequently, in the right-hand member of Equation 1, we obtain $f(\alpha_1', \alpha_2', \ldots, \alpha_n')$.

We have obtained

$$f(\alpha_1', \alpha_2', \ldots, \alpha_n') = f(\alpha_1', \alpha_2', \ldots, \alpha_n')$$

for an arbitrary n-tuple $(\alpha_1', \alpha_2', \ldots, \alpha_n')$ of values of the variables. This completes the proof of Equation 1.

Since $f(\alpha_1, \alpha_2, \ldots, \alpha_n)$ is either 1 or 0, the disjunction in the right-hand member of Equation 1 contains only the terms for which $f(\alpha_1, \alpha_2, \ldots, \alpha_n) = 1$; that is, Equation 1 can be written

$$f(X_1, X_2, \ldots, X_n) = \bigvee_1 X_1^{\alpha_1} X_2^{\alpha_2} \ldots X_n^{\alpha_n}.$$

We have obtained the perfect disjunctive normal form of the function $f(X_1, X_2, \ldots, X_n)$.

Obviously, if the function is identically equal to 0, it cannot be represented as a perfect disjunctive normal form. On the other hand, if the function is identically equal to 1, its perfect disjunctive normal form contains all 2^n terms corresponding to all possible n-tuples of values of the arguments. For example, the function of two variables that is identically equal to 1 is represented as a perfect disjunctive normal form as follows:

$$1 = \bar{X}\bar{Y} \vee \bar{X}Y \vee X\bar{Y} \vee XY.$$

Analogously, one can prove that an arbitrary function of n variables that is not identically equal to 1 can be represented as a perfect conjunctive normal form:

$$f(X_1, X_2, \ldots, X_n) = \bigwedge_0 (X_1^{\bar{\alpha}_1} \vee X_2^{\bar{\alpha}_2} \vee \ldots \vee X_n^{\bar{\alpha}_n}).$$

If the function is identically equal to 0, then its perfect conjunctive normal form contains all 2^n terms corresponding to all possible sets of values of the arguments. For example, 0 as a function of two variables is represented as a perfect conjunctive normal form as follows:

$$0 = (X \vee Y)(X \vee \bar{Y})(\bar{X} \vee Y)(\bar{X} \vee \bar{Y}).$$

EXERCISES

1.28. We have proved that a function not identically 0 can be represented as a perfect disjunctive normal form. Go through the analogous proof that a function not identically 1 can be represented as a perfect conjunctive normal form.

1.29. Starting with the truth tables defining the implication $X \Rightarrow Y$ and the strict disjunction $X \dot{\vee} Y$, construct the perfect disjunctive and conjunctive normal forms of these functions.

1.30. Define with the aid of tables all possible functions of two variables. It will be expedient to set up a common table:

X	Y	f_1	f_2	...	f_{16}
0	0				
0	1				
1	0				
1	1				

For these functions, write the perfect disjunctive and conjunctive normal forms. Simplify the formulas obtained.

4.2. The expression defining a function with the aid of one of the normal forms is not in general the shortest expression for this function. For example, the perfect disjunctive normal form of the function introduced on p. 74,

$$f(X, Y, Z) = \bar{X}YZ \vee X\bar{Y}Z \vee XY\bar{Z} \vee XYZ, \tag{a}$$

can be simplified in various ways. For example, by grouping the last term with the first or second or third, we obtain the following three forms for this function:

$$f(X, Y, Z) = YZ \vee X\bar{Y}Z \vee XY\bar{Z}, \tag{b}$$
$$f(X, Y, Z) = \bar{X}YZ \vee XZ \vee XY\bar{Z}, \tag{c}$$
$$f(X, Y, Z) = \bar{X}YZ \vee X\bar{Y}Z \vee XY. \tag{d}$$

These are known as disjunctive normal forms (dnf) for the function.

As one can see, a single function has many disjunctive normal forms. The perfect disjunctive normal form is distinguished from the others by the fact that its terms are elementary conjunctions all of the same rank (equal to the number of arguments).

In connection with applications of the algebra of propositions, in particular, to the theory of electric networks, the problem arises of finding the minimal (i.e., containing the fewest number of letters) forms among all the disjunctive or conjunctive normal forms of a given function. (This is the *minimization problem*.)

Although forms b, c, and d contain fewer letters than the perfect disjunctive normal form a of the function, they are not *minimal* disjunctive normal forms of that function. In fact, if we add disjunctively to the perfect disjunctive normal form of this function two more conjunctions XYZ [which, by virtue of the idempotency law (8) leads to a function equivalent to the original one] and use laws 2 and 6, we obtain

$$f(X, Y, Z) = YZ(\bar{X} \vee X) \vee XZ(\bar{Y} \vee Y) \vee XY(\bar{Z} \vee Z).$$

If we now use laws 14 and 11, we obtain a minimal* disjunctive normal form of the given function:

$$f(X, Y, Z) = YZ \vee XZ \vee XY.$$

However, we have used here an artificial device that may not be suitable for another function.

The problems of minimizing Boolean functions, which are of exceptional importance for applications of mathematical logic in technology (the design of electronic computing machines and other automatic devices), have been subject to intense development in recent years. A consideration of this problem would, however, take us beyond the framework of the present book.†

4.3. An arbitrary formula in the algebra of propositions can be reduced to conjunctive normal form. For this it is necessary:

1. to express all operations the symbols for which appear in the formula in terms of disjunction, conjunction, and negation (on the basis of laws 18 and 19);

---

* That is, containing the smallest possible number of letters. Of course, the fact that it is actually minimal still needs to be proved but, for want of space, we shall not do this here.
† The reader who wishes to know more about the miminization problem should consult Hohn [4], Mendelson [7], and the following three papers by W. V. Quine in the *American Mathematical Monthly*.
(1) The problem of simplifying truth functions, 59, 1952, 521–531.
(2) A way to simplify truth functions, 62, 1955, 627–631.
(3) On cores and prime implicants of truth functions, 66, 1959, 755–760.

2. to reduce the symbols for negation to elementary propositions (on the basis of laws 16 and 17);

3. to use the laws of dual negation, associativity, and commutativity of disjunction and conjunction and distributivity of disjunction with respect to conjunction (laws 1–5, 7). We have encountered all these transformations several times in various exercises.

The reduction of a formula to conjunctive normal form can be used as a decision procedure in answering the question as to whether or not this formula is a tautology. Specifically, if every term of the conjunctive normal form is an elementary disjunction containing at least one variable together with its negation, that is, if it is of the form

$$X_1^{\alpha_1} \vee X_2^{\alpha_2} \vee \ldots \vee X_i^{\alpha_i} \vee X_i^{\bar{\alpha}_i} \vee \ldots \vee X_k^{\alpha_k},$$

then this disjunction has the value 1 and, consequently, the conjunction as a whole has the value 1 for all combinations of values of the variables; that is, it is a tautology. On the other hand, if at least one term of the conjunctive normal form fails to contain at least one variable together with its negation, then the conjunction is not a tautology since there then exists a combination of values of the variables for which this term and hence the conjunction as a whole has the value 0. For example, if

$$X_1^{\alpha_1} \vee X_2^{\alpha_2} \vee \ldots \vee X_n^{\alpha_n}$$

is such a term of the conjunction, then, for the n-tuple $(\bar{\alpha}_1, \bar{\alpha}_2, \ldots, \bar{\alpha}_n)$ the conjunction assumes the value 0; that is, it is not a tautology. This decision procedure is sometimes more convenient in practice than setting up a truth table.

As an example, let us see whether the formula

$$(X \Rightarrow Y) \Rightarrow ((X \Rightarrow Z) \Rightarrow (X \Rightarrow YZ))$$

(Exercise 1.22d on p. 62) is a tautology or not by reducing it to conjunctive normal form:
equiv $\overline{\overline{X} \vee Y} \vee \overline{\overline{X} \vee Z} \vee \overline{X} \vee YZ$;
equiv $X\overline{Y} \vee X\overline{Z} \vee \overline{X} \vee YZ$;

equiv $X(\bar{Y} \lor \bar{Z}) \lor \bar{X} \lor YZ$;
equiv $(X \lor \bar{X})(\bar{Y} \lor \bar{Z} \lor \bar{X}) \lor YZ$;
equiv $(X \lor \bar{X} \lor YZ)(\bar{Y} \lor \bar{Z} \lor \bar{X} \lor YZ)$;
equiv $(X \lor \bar{X} \lor Y)(X \lor \bar{X} \lor Z)(\bar{Y} \lor \bar{Z} \lor \bar{X} \lor Y)$
$(\bar{Y} \lor \bar{Z} \lor \bar{X} \lor Z)$.

We have obtained a conjunction every term of which is an elementary disjunction containing a variable together with its negation. Thus, this conjunction is a tautology and, consequently, Formula 1.22d, which is equivalent to it, is also a tautology.

EXERCISE

1.31. Show that formulas a, b, c, e, and f of Exercise 1.22 are tautologies by reducing them to conjunctive normal form.

5. APPLICATION OF THE ALGEBRA OF PROPOSITIONS TO THE SYNTHESIS AND ANALYSIS OF DISCRETE-ACTION NETWORKS

5.1. Automatic devices can be divided into discrete- and continuous-action devices.

The operation of discrete devices, for example, digital computers, is characterized by a stepwise variation between a finite number of states; the operation of continuous devices, for example, simulating (analogue) machines, is characterized by a continuous variation in the states.

The physical nature of a device is determined by the electronic, mechanical, and other characteristics of its parts. In contrast, the *functional* characteristics of these parts (which take into account the purpose of each part but not the means used for carrying out that purpose) and the way in which they are put together constitute the logical structure of the device.

By the logical synthesis of a discrete-action network, we mean the determination of the logical structure of a discrete device from given conditions concerning its operation. By logical analysis, we mean the opposite problem, that is, the problem of determining the conditions of operation of a device from its known logical structure.

In what follows, we shall omit the word "logical" and simply speak of the synthesis and analysis of a network.

We shall consider only the simplest cases, those in which the elements of the device can have only two states; that is, they operate according to a yes-no principle (or a closed-open principle), just as propositions can assume only two values, namely, true or false. Various electrical switches that surround us at work and in our homes are examples of elements of the yes-no type.

The fundamental similarity between elements of this type and propositions, which consists in the fact that both can assume only two states (two values), serves as the foundation for the application of the algebra of propositions to the synthesis and analysis of networks consisting of such elements.

The idea of the possibility of such application was proposed as far back as 1910 by the physicist P. Ehrenfest. However, rigorous proofs of the possibility and the procedure for applying the algebra of propositions to the synthesis and analysis of electrical circuits were first developed in the thirties by the Soviet scholar V. I. Shestakov and the American scholar C. E. Shannon.

Below, we shall consider the simplest contact and contactless schemes to the synthesis and analysis of which the apparatus of the algebra of propositions can be applied.

5.2. The elements from which contact networks are constructed are electric contacts with two states: closed (in which the contact closes the circuit and allows a current to pass through it) and open (in which the circuit is open and current cannot pass through it). Here, we shall not concern ourselves with the question as to whether contacts can be changed from one state to another with the aid of an electromagnetic relay or manually or by some other means.

Closing and opening contacts are used in the schemes. The first of these close the circuit in an operating state and open it in a nonoperating state; the second does just the opposite.

In the drawings, the contacts are shown in a nonoperating state (in Figure 1a, a closing contact is shown; in Figure 1b, an opening contact is shown).

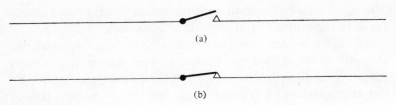

Figure 1

Contacts that are closed or opened by the same governing element (a relay, a switch, etc.) are indicated by the same capital letter.

The application of the algebra of propositions to the synthesis and analysis of contact networks is based on the possibility of a different interpretation from that considered above, a nonlogical interpretation of abstract Boolean algebra in terms of electrical network theory.

To make this interpretation, we need only the following initial dictionary:

The Language of Abstract Boolean Algebra	The Language of the Algebra of Propositions	The Language of the Algebra of Contact Networks
$A, B, C, \ldots, X, Y, Z$, ... are variables	... are propositions each of which can be either true or false	... are contacts each of which can be either closed or open
1 is the value of the variable	T (the proposition is true)	1 (the contact is closed)
0 is the value of the variable	F (the proposition is false)	0 (the contact is open)

On the basis of this dictionary, it is easy to make clear the "meanings" of disjunction, conjunction, and negation under the new interpretation:

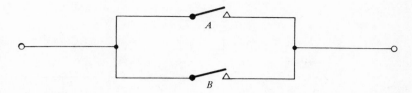

Figure 2

1. Corresponding to the disjunction $A \vee B$ is a network consisting of contacts A and B connected in parallel. It is closed if and only if at least one of the contacts A or B is closed (see Figure 2). Under one interpretation of the symbols A, B, 0, and 1, the table

A	B	$A \vee B$
0	0	0
0	1	1
1	0	1
1	1	1

shows how the truth value of the disjunction of the two propositions A and B depends on the truth values of these two propositions. Under another interpretation, this table determines the state (closed or open) of a network consisting of the two contacts A and B connected in parallel as a function of the states of these two contacts.

2. The conjunction AB corresponds to a network consisting of contacts A and B connected in series. It is closed if and only if both of these two contacts are closed (see Figure 3).

3. The negation $\bar{A}$ corresponds to a contact that is closed when A is open and is open when A is closed (see Figure 4).

Figure 3

SEC. 5. APPLICATIONS 85

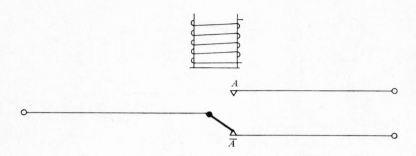

Figure 4

Since every function in the algebra of propositions can be expressed in either disjunctive or conjunctive normal form, that is, with the aid of the symbols ¯, ∧, and ∨ (for negation, conjunction, and disjunction, respectively (see Section 4)), the correspondence described above assigns to each function in the algebra of propositions a contact network made up of closed and open contacts with the aid of series and parallel connections. Such networks are called series-parallel networks, or P-networks, or networks of the class P.

Every function in the algebra of propositions can be realized with the aid of a P-network. The network corresponding to a given function is called its network realization.

Obviously, to every network in the class P there corresponds a function expressed by a formula built from variables and their negations with the aid of the symbols for disjunction and conjunction. This formula is called the **structural formula** of the network. This correspondence is the basis for the application of the algebra of propositions to the analysis, simplification, and synthesis of contact networks.

The analysis of a network, that is, the determination of the conditions of operation (closing and opening) of the given network, reduces to determination of the values of the structural formula corresponding to this network under all possible combinations of values of the variables. Simplification of a given network reduces to simplification of its structural formula.

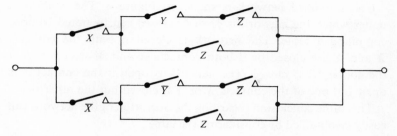

Figure 5

Synthesis of a network from given conditions for its operation reduces to setting up its structural formula from these conditions (given in tabular form) and to the construction of the network corresponding to that formula.

Let us give some examples of the analysis, simplification, and synthesis of contact networks.

Problem 1. Analyze and, if possible, simplify the network shown in Figure 5.

From the given network, one can easily write its structural formula:

$f(X, Y, Z) = X(Y\bar{Z} \vee Z) \vee \bar{X}(\bar{Y}\bar{Z} \vee Z).$

Let us simplify this formula and then make an analysis of the simplified network (equivalent to the given one with regard to the passage of a current, for any state of the contacts):

$$X(Y\bar{Z} \vee Z) \vee \bar{X}(\bar{Y}\bar{Z} \vee Z) = X(Y \vee Z) \vee \bar{X}(\bar{Y} \vee Z);$$
$$= XY \vee XZ \vee \bar{X}\bar{Y} \vee \bar{X}Z;$$
$$= XY \vee \bar{X}\bar{Y} \vee Z.$$

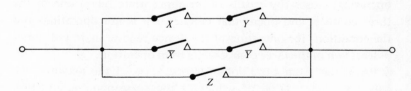

Figure 6

SEC. 5. APPLICATIONS

The simplified network is shown in Figure 6. The conditions under which the network is open or closed can be noted without compiling a table. The network is closed if both the contacts X and Y are closed or if both contacts X and Y are open or if the contact Z is closed. The network is open if the contact Z is open and one of the contacts X or Y is closed but the other open.

This same conclusion regarding the operation of the network can easily be obtained by considering the table

X	Y	Z	XY	$\bar{X}$	$\bar{Y}$	$\bar{X}\bar{Y}$	$XY \vee \bar{X}\bar{Y} \vee Z$
0	0	0	0	1	1	1	1
0	0	1	0	1	1	1	1
0	1	0	0	1	0	0	0
0	1	1	0	1	0	0	1
1	0	0	0	0	0	0	0
1	0	1	0	0	1	1	1
1	1	0	1	0	0	0	1
1	1	1	1	0	0	0	1

Let us give an example of the synthesis of a network.

Problem 2. From three contacts A, B, and C, construct a network with one input and one output in such a way that a signal will appear (for example, a light will turn on) at the output if at least two of the three contacts are closed.

An application of such a network is the control of a device consisting of three separate components each of which, when it is operating, closes (by means of an appropriate relay) one of the three contacts and opens that contact if it is not operating, and the conditions for operation of the device require that at all times at least two of the three components be in operation.

From the given conditions for operation of the network, we can construct a table of values of the corresponding function $f(A, B, C)$. Beginning with the table of the function

A	B	C	f(A, B, C)
0	0	0	0
0	0	1	0
0	1	0	0
0	1	1	1
1	0	0	0
1	0	1	1
1	1	0	1
1	1	1	1

we can write its perfect disjunctive normal form

$$f(A, B, C) = \bar{A}BC \vee A\bar{B}C \vee AB\bar{C} \vee ABC.$$

In Section 4, we found the minimal disjunctive form of this function:

$$f(A, B, C) = BC \vee AC \vee AB.$$

We note, however, that the minimal *disjunctive* normal form may not be minimal among all possible forms of this function. Thus, in the present case, we can decrease further the number of letters by one letter:

$$f(A, B, C) = BC \vee A(C \vee B).$$

We obtain the network shown in Figure 7.

5.3. Present-day electronic computing machines attain astounding speeds with the aid of contactless networks that operate

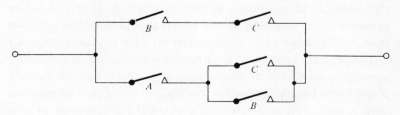

Figure 7

thousands of times as fast as the corresponding relay-contact networks. These contactless networks use vacuum tubes (diodes, triodes, pentodes, etc.) or semiconducting devices that carry out the basic logical operations (negation, disjunction, and conjunction).

We shall not go into the structure or physical bases of these devices, known as **functional elements**. Let us denote them as follows:

NOT denotes a device performing negation;
OR denotes a device performing disjunction;
AND denotes a device performing conjunction.

Of these functional elements, we know only the following:

1. The element NOT has one input and one output. The signal appears at the output when there is no signal at the input and a signal fails to appear at the output when there is a signal at the input.

2. The element OR has two or more inputs and one output. A signal appears at the output if and only if a signal is applied at one or more of the inputs.

3. The element AND has two or more inputs and one output. A signal appears at the output if and only if signals are applied at all the inputs.

We shall use only these functional characteristics of the elements NOT, OR, and AND when we are solving the problem of synthesis of a network consisting of these elements.

As an example, let us look at the problem of synthesizing a single-bit binary adder with three inputs. (An adder carrying out the addition of multidigit binary numbers is simply a series combination of single-bit binary adders that perform the addition in each bit and carry to a higher-order bit if the situation requires.)

The problem consists in constructing with the aid of functional elements a network with three inputs A, B, and C and two outputs S and P (see Figure 8) in such a way that, when signals representing binary digits (the addends of a given digit) are applied at two inputs (let us say, A and B) and a signal representing the value of the carry from the preceding digit is applied at the input C, one obtains at the output S the value of the sum in the given digit and at the output P the value of the carry to the next higher digit.

A, B, and C, which assume the values 0 and 1, are Boolean (two-valued) variables and S and P are Boolean functions of A, B, and C:

$S = S(A, B, C)$,
$P = P(A, B, C)$.

If we compile a table for addition in the binary system of numeration and interpret addition as disjunction, multiplication as conjunction, and negation as replacing 1 with 0 and vice versa, we obtain the following perfect disjunctive normal forms of these functions:

$P(A, B, C) = \bar{A}BC \vee A\bar{B}C \vee AB\bar{C} \vee ABC$;
$S(A, B, C) = \bar{A}\bar{B}C \vee \bar{A}B\bar{C} \vee \bar{A}\bar{B}\bar{C} \vee ABC$.

For the function $P(A, B, C)$, we already know the minimal disjunctive form (see Exercise 4.2, p. 79):

$P(A, B, C) = BC \vee AC \vee AB$.

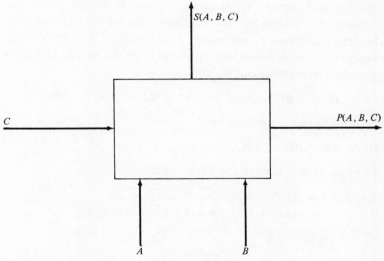

Figure 8

SEC. 5. APPLICATIONS 91

A	B	C	P	S
0	0	0	0	0
0	0	1	0	1
0	1	0	0	1
0	1	1	1	0
1	0	0	0	1
1	0	1	1	0
1	1	0	1	0
1	1	1	1	1

Since the network is constructed from functional elements, the problem arises as to how to simplify the formula in such a way that it will contain as few symbols for operations as possible. Obviously, the smallest number of operations can be obtained by reducing the formula to a form that will have the negation symbols cover the longest expressions possible. Thus, by using de Morgan's laws 16 and 17, we obtain

$$S(A, B, C) = ABC \vee \overline{(A \vee B \vee \overline{C})(A \vee \overline{B} \vee C)(\overline{A} \vee B \vee C)}.$$

To simplify the expression, let us use the distributivity of conjunction with respect to disjunction (Law 6) to get rid of the parentheses. This can be done as follows: We take each letter from the first parenthetical expression and combine it with nonzero conjunctions, each containing one letter from the second and third parenthetical expressions. We obtain

$$AB \vee AC \vee A\overline{B}C \vee ABC \vee \overline{A}BC \vee BC \vee AB\overline{C} \vee \overline{A}\overline{B}\overline{C};$$

and after further simplification

$$AC \vee AB \vee BC \vee \overline{A}\overline{B}\overline{C}.$$

If we again apply laws 16 and 17, we obtain

$$S(A, B, C) = ABC \vee \overline{\overline{A}\overline{B}\overline{C} \vee AB \vee AC \vee BC}$$
$$= ABC \vee (A \vee B \vee C)\overline{AB \vee AC \vee BC}.$$

Thus, by applying

$$S = ABC \vee (A \vee B \vee C)\overline{AB \vee AC \vee BC}$$

or

$$S = ABC \vee (A \vee B \vee C)\bar{P} \text{ and } P = AB \vee AC \vee BC,$$

we obtain the network of a single-bit binary adder with three inputs shown in Figure 9.

EXERCISES

1.32. From the contacts A, B, C, $\bar{A}$, $\bar{B}$, and $\bar{C}$, construct a network that will be closed if and only if at least two of the three contacts A, B, C are closed.

1.33. With the aid of functional elements carrying out the basic logical operations, construct a network with three inputs and one output so that there will be a signal at the output if and only if signals are applied at two inputs.

1.34. With the aid of functional elements that carry out the basic logical operations, construct a network of a single-bit binary adder with two inputs. How should we connect two such adders in order to obtain a single-bit adder with three inputs?

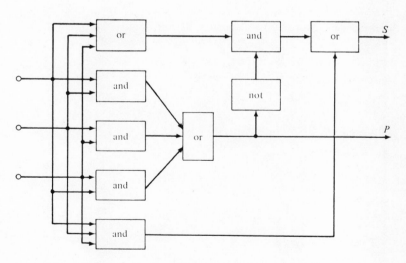

Figure 9

1.35. With the aid of four functional elements realizing the basic logical operations, construct a network with two inputs and two outputs in such a way that at one output a signal will appear if and only if a signal is applied at one or both of the inputs and at the other output a signal will appear when a signal is applied at only one of the inputs.

1.36. Construct contact and contactless networks (consisting of functional elements) that perform a transformation of single-digit octonary numbers into binary numbers.

1.37. Construct a network realization of the functions defined by the formulas

a. $(X \Rightarrow Y)(Y \Rightarrow Z)$;
b. $(X \Rightarrow Y)(Y \Rightarrow Z) \Rightarrow (X \Rightarrow Z)$.

2
THE PROPOSITIONAL CALCULUS

1. THE AXIOMATIC METHOD. THE CONSTRUCTION OF FORMALIZED LANGUAGES

1.1. Although we cannot go into the history of the axiomatic method here,* we mention two characteristic features of it, which will be important for our subsequent exposition:

a. The axiomatic system of Euclidean geometry constructed by Hilbert in his *Grundlagen der Geometrie* (1899) differs from the system described in the *Elements of Euclid* (third century B.C.) not only in the completeness of the system of axioms for a rigorous logical development of all of geometry but in its completely new understanding of axiomatic theory.

Prior to Hilbert's time, mathematicians, in numerous attempts at axiomatic construction of geometry, had developed a theory as a description of some specific system of objects. The axioms were treated as "obvious" propositions regarding these objects, and the problem of axiomatization was reduced to the logical deduction of all the propositions in the theory from the original ones (the axioms).

Under such a contensive ("meaningful") view, the axioms were considered as contensive propositions (having a real meaning) regarding the objects of the theory. The expression "logical deduction" was assumed (although without sufficient justification) to be unambiguous and easily understood.

The work of Hilbert is connected with the transition from the contensive to the formal understanding of axiomatic theory.

Hilbert begins his axiomatic construction of geometry with the words†

Let us consider three distinct systems of things. The things composing the first system, we will call *points* and designate them by the letters A, B, C, ...; those of the second, we will call *straight lines* and designate them by the letters a, b, c, ...; and those of the third system, we will call *planes* and designate them by the Greek letters α, β, γ, ...

In the set of all these objects, one introduces certain relation-

---

* See the literature recommended in the footnote on p. 5.
† D. Hilbert, *The Foundations of Geometry*, Chicago, Open Court, 1902, p. 3 (translation of *Grundlagen der Geometrie*).

ships, indicated by such words as "incident," "belongs," "between," "congruent," etc.

In the development of the theory describing the structure of this set of objects (points, lines, planes) and the relationships among those objects, we not only do not consider the "nature" of the objects but do not even consider the "meaning" of the *relationships* between them. In it, we use only certain formal properties of these relationships, which constitute the content of the geometrical axioms.

Under Hilbert's approach, the terms "point," "line," "plane," "incident," "between," "congruent," etc., have no intuitively understandable meaning other than what is explicitly indicated in the axioms.

The transition from the contensive to the formal understanding of axiomatic theory can be vividly (although not precisely) described as the transition from *axiomatization of content* to *axiomatization of form* of the language in which this content is expressed. Since a given form can be given several distinct meanings, this also implies a transition from a theory describing one specific domain of objects to a theory describing a class of domains of objects. In general, these objects are quite different in nature and the meanings of the relationships connecting them are also different though these domains have the same formal structure.

These object domains are different concrete (contensive) *models* of a single abstract theory. We say also that different *interpretations* of the theory are possible with the aid of objects and relationships pertaining to different models.* This makes it possible to apply a single abstract mathematical theory to various domains of objects.

b. The language of a scientific theory consists of two component parts: (1) the strictly technical language in that theory, to which

---

* The abstract theory can also be interpreted in terms of another abstract theory. An example is the arithmetic interpretation of Euclidean geometry whereby a point is interpreted as a pair of real numbers (x, y) and a line as ratios $a:b:c$ with a and b not simultaneously equal to 0. Also, the statement "the point (x, y) is incident to the line $a:b:c$" is interpreted as the equation $ax + by + c = 0$, etc. For more detail, see the literature cited in the footnotes on pages 5 and 73.

belong the terms and symbols denoting the objects (together with their properties and relationships) that are studied in it and (2) a logical language with the aid of which we can use the terms in the technical language to construct the propositions of the theory and derive certain propositions from others.

In Hilbert's axiomatic system, only a portion of the language of geometric theory (purely geometric language) was subjected to axiomatization. The other portion, logic, which was applied in the development of geometric theory, remained unformalized and unaxiomatized.

One of the directions of later development of the axiomatic method (also developed primarily by Hilbert and his school) was the extension of the formalization and axiomatization to the logic by means of which the axiomatic theory is developed, and to the language (modes of expression) of that logic. The axiomatic method came to be regarded as a method of constructing formalized languages. The concept of a formal (or deductive) system or calculus arose as a natural precise version of the concept of axiomatic theory.

A formal system includes, together with the system of axioms of the theory being axiomatized, a system of logical axioms and deduction rules. A logical calculus used as the logical language of an axiomatic theory must be capable of expressing the axioms of that theory in a way that will be suitable for the derivation from the axioms of all the consequences in which we are interested, and it must provide us with adequate tools for such a derivation.

The propositional calculus is in itself insufficient as a logical language for axiomatic theories but is a component part of the majority of other rich logical calculi that fulfill this purpose successfully.

1.2. The formal part of the formalized language, the calculus, is constructed according to the following scheme:

a. First of all, we have a list of symbols, an alphabet for the calculus. (We recall that we do not associate with the symbols of the alphabet any particular meaning. These are simply signs that we know how to recognize and distinguish by their appearance.)

b. From the symbols of the alphabet, we construct the formulas of the calculus, the basic linguistic formations. The set of formulas of the calculus is determined, for example, by a specification of the atomic (elementary) formulas and formation rules with the aid of which one can construct new formulas from old ones.

c. In the set of formulas of the calculus, we single out by some means or other the subset of derived formulas. This can be done, for example, by indicating the initial formulas (axioms) and deduction rules with the aid of which we can obtain new derived formulas from the derived formulas that we already have.

It should be noted that the rules of formation and the rules of deduction are assertions regarding the formulas of the calculus, formulated in the metalanguage (cf. footnote on p. 48), that is, in that language in which we are studying the structure of the given calculus.

A calculus becomes a *formalized language* when, to every formal expression, we give an *interpretation* (see p. 97).

Let us give an example of the construction, according to the outline indicated above, of a simple calculus whose alphabet consists of only three symbols.

A. The calculus alphabet $\{|, +, =\}$

B. The calculus formulas
   a. The atomic formulas: $|$
   b. The rules of formation of formulas:
   Formation rule 1. If φ is a formula, then $\varphi|$ is a formula.*
   Formation rule 2. If φ_1 and φ_2 are formulas, then $\varphi_1 + \varphi_2$ is a formula.
   Formation rule 3. If φ_1 and φ_2 are formulas, then $\varphi_1 = \varphi_2$ is a formula.

---

* The symbols φ, φ_1, φ_2, φ_3, ..., are not part of the alphabet of our calculus (see A above). These are metalinguistic symbols that are used to indicate the fact that, in place of any one of them, we may have in mind an arbitrary formula of the given calculus.

C. Derived formulas
  a. The initial formulas (axioms):
    1. $| + | = | |$.
  b. Rules of inference:
    Rule 1. If $\varphi_1 + | = \varphi_2$ is a derived formula, then $\varphi_1 | + | = \varphi_2 |$ is a derived formula.
    Rule 2. If $\varphi_1 + \varphi_2 = \varphi_3$ is a derived formula, then $\varphi_1 + \varphi_2 | = \varphi_3 |$ is a derived formula.

With the aid of the atomic formula and the formation rules, we can determine, for an arbitrary finite sequence of symbols in the alphabet of the calculus, whether it is a formula in this calculus or not.

Let us take as an example the following finite sequence of symbols from the alphabet of the calculus: $| | + | | = | | | |$.

If we can find a finite sequence of formulas each of which either is an atomic formula or is obtained from the preceding formulas in accordance with some formation rule and the final formula is $| | + | | = | | | |$, then this sequence of symbols is a formula. On the other hand, if we can show that it is impossible to find such a sequence of formulas, then the given sequence of symbols is not a formula.

Below, we give a finite sequence of formulas proving that $| | + | | = | | | |$ is a formula. In each row, the parenthetical expressions indicate from which of the preceding formulas or in accordance with what formation rule the formula in that row is obtained. If it is obtained from the directly preceding formula, only the formation rule according to which it is obtained will be indicated.

1. $|$ (Atomic formula)
2. $| |$ (Formation rule 1)
3. $| | + | |$ (Formation rule 2)
4. $| | |$ (2, Formation rule 1)
5. $| | | |$ (Formation rule 1)
6. $| | + | | = | | | |$ (3, 5, Formation rule 3)

One can easily show that the formula $|| + || = ||||$ is a derived formula.

Proof: This must be a finite sequence of formulas each of which either is an initial derived formula (axiom) or is obtained from the preceding ones in accordance with one of the rules of inference, and the final formula must be $|| + || = ||||$.

1. $| + | = ||$ (A)
2. $|| + | = |||$ (Rule of inference 1)
3. $|| + || = ||||$ (Rule of inference 2)

If we understand by $|, ||, |||, ||||, \ldots$, the natural numbers 1, 2, 3, 4, ..., if we understand by + the symbol for ordinary addition, and if we understand by = the symbol for equality, we obtain a *model* of the above-described calculus in terms of ordinary arithmetic, in which "$2 + 2 = 4$" is a true formula.

EXERCISES

2.38. Expand the above-described calculus as follows:

1. Add to the alphabet the symbol $\times$.
2. Admit formation rule 4:
   If φ_1 and φ_2 are formulas, then $\varphi_1 \times \varphi_2$ is a formula.
3. Admit the new axiom: $| \times | = |$.
4. Admit the following two rules of inference:
   Rule 3. If $\varphi_1 \times | = \varphi_2$ is a derived formula, then $\varphi_1| \times | = \varphi_2|$ is a derived formula.
   Rule 4. If $\varphi_1 \times \varphi_2 = \varphi_3$ and $\varphi_3 + \varphi_1 = \varphi_4$ are derived formulas, then $\varphi_1 \times \varphi_2| = \varphi_4$ is a derived formula.

In the (expanded) calculus obtained, prove

a. $|| \times || = ||||$ is a formula.
b. $|| \times || = ||||$ is a derived formula.

2. CONSTRUCTION OF A PROPOSITIONAL CALCULUS (ALPHABET, FORMULAS, DERIVED FORMULAS)

Let us consider a logical calculus (an axiomatic logical system) one of the possible interpretations of which can be constructed in terms of the meaningful algebra of propositions. This calculus is called the **propositional calculus**.

Although, in describing the construction of this calculus, we use familiar terminology and symbols, we shall not associate with them the meaning that they have in the meaningful algebra of propositions (see Chapter 1) nor any other specific meaning.

A. The alphabet of the propositional calculus consists of:

1. Capital Latin letters A, B, C, ..., X, Y, Z (with or without subscripts). These are called **variables** or **propositional variables**. No specific meaning should be ascribed to the term "propositional."
2. The symbols $^{-}$, $\wedge$, $\vee$, and $\Rightarrow$. These are called respectively the symbols for "negation," "conjunction," "disjunction," and "implication." Again, no specific meaning should be given to these terms.
3. The pair of symbols ( ).

There are no other symbols in the alphabet of the propositional calculus.

For example, the Greek letters φ_1, ψ, ... (with or without subscripts), which we shall use to shorten the notation in certain constructions made from the symbols of the alphabet, do not themselves belong to this alphabet. They are metalinguistic symbols.

Instead of four symbols (for negation, conjunction, disjunction, and implication), we might include in the alphabet of the propositional calculus only two symbols. For example, in the axiomatic system of the propositional calculus constructed by Frege, one takes as initial symbols the symbols for implication and negation. In the system of Russell and Whitehead, one takes the symbols for negation and disjunction. In one of Hilbert's systems, one takes the symbols for disjunction and implication. In such cases, the other two symbols are used for shortening the notation for certain constructions from the symbols of the alphabet.

B. The formulas in the propositional calculus are finite sequences (rows) of symbols belonging to the alphabet of that calculus, which satisfy the following definition:

a. The atomic (elementary) formulas: $A, B, C, \ldots, X, Y, Z, \ldots$, and $X_1, X_2, X_3, \ldots$, that is, the variables.
b. The rules of formation of formulas:
  Formation rule 1: If φ is a formula, then $\bar{\varphi}$ is a formula.
  Formation rule 2: If φ_1 and φ_2 are formulas, then $(\varphi_1 \wedge \varphi_2)$ is a formula.
  Formation rule 3: If φ_1 and φ_2 are formulas, then $(\varphi_1 \vee \varphi_2)$ is a formula.
  Formation rule 4: If φ_1 and φ_2 are formulas, then $(\varphi_1 \Rightarrow \varphi_2)$ is a formula.

c. There are no formulas other than the formulas enumerated in a or obtained from these with the aid of the rules enumerated in b.

Let us use this definition to show, for example, that the following finite sequence of symbols of the alphabet

$$(((A \wedge B) \vee C) \Rightarrow ((\bar{A} \vee B) \Rightarrow \bar{C})) \tag{1}$$

is a formula.

1. A (Atomic formula).
2. B (Atomic formula).
3. C (Atomic formula).
4. $\bar{A}$ (1, Formation rule 1).
5. $\bar{C}$ (3, Formation rule 1).
6. $(A \wedge B)$ (1, 2, Formation rule 2).
7. $((A \wedge B) \vee C)$ (6, 3, Formation rule 3).
8. $(\bar{A} \vee B)$ (4, 2, Formation rule 3).
9. $((\bar{A} \vee B) \Rightarrow \bar{C})$ (8, 5, Formation rule 4).
10. $(((A \wedge B) \vee C) \Rightarrow ((\bar{A} \vee B) \Rightarrow \bar{C}))$ (7, 9, Formation rule 4).

Formulas 1–9, which are formed in the process of constructing Formula 10, are called the parts of Formula 10.

Let us now determine whether the sequence of symbols

$$(X \vee Y) \Rightarrow Z) \tag{2}$$

is a formula or not.

We note that the atomic formulas a contain no parentheses and that the parentheses appearing in the formulas obtained with the aid of formation rules 2–4 always come in pairs (consisting of one left-hand parenthesis and one right-hand parenthesis). Therefore, taking c into account, we can assert that any formula of the propositional calculus contains an even number of parentheses, in fact, just as many left-hand as right-hand parentheses. From this it follows that Sequence 2 is not a formula.

Similarly, the sequence of symbols

$$(X \wedge) \tag{3}$$

is not a formula. This is true because, in the formation of formulas with the aid of the conjunction symbol (Formation rule 2), there must appear a formula both to the left and to the right of this symbol. In Sequence 3, the formula X (listed under a) appears only to the left of the conjunction symbol and no formula appears to the right of it. Therefore, in view of c, Sequence 3 is not a formula.

The sequence of symbols

$$(X \Rightarrow Y) \vee Z(\tag{4}$$

is not a formula since no formula containing parentheses (Formation rules 2–4) can have a left-hand parenthesis at its right end.

In order to reduce the use of parentheses in complicated formulas, let us agree to drop the external parentheses, that is, those parentheses that include all the other symbols appearing in the formula. Furthermore, just as in the algebra of propositions, we shall treat the symbol $\wedge$ as being a closer connective than the symbol $\vee$ or the symbol $\Rightarrow$ and we shall assume that the symbol $\vee$ is more closely connecting than the symbol $\Rightarrow$. For greater compactness in writing the formulas, let us also agree to omit the symbol for conjunction, that is, to write simply XY instead of $X \wedge Y$.

With these conventions, Formula 1 may be written

$$AB \vee C \Rightarrow (\overline{A} \vee B \Rightarrow \overline{C}).$$

C. From the set of formulas of the propositional calculus, we must pick out the subset of derived formulas. The class of derived formulas is defined as the set of all possible formulas that can be obtained from certain formulas taken as initial derived formulas (axioms) by means of certain special rules (rules of inference).

As a basis for the propositional calculus, just as for any axiomatically constructed theory, one can take different systems of axioms that are equivalent to each other in the sense that the class of derived formulas determined by these systems is the same for one system as for another.*

We take as our basis for the propositional calculus the system of axioms proposed by P. S. Novikov [8]. This system consists of eleven axioms divided into four groups. The axioms of the first group contain only the symbol for implication. The axioms of the second group contain the symbols for implication and conjunction. The axioms of the third group contain the symbols for implication and disjunction. The axioms of the fourth group contain the symbols for implication and negation.

AXIOMS
I.

I.1. $A \Rightarrow (B \Rightarrow A)$;
I.2. $(A \Rightarrow (B \Rightarrow C)) \Rightarrow ((A \Rightarrow B) \Rightarrow (A \Rightarrow C))$.

II.

II.1. $AB \Rightarrow A$;
II.2. $AB \Rightarrow B$;
II.3. $(A \Rightarrow B) \Rightarrow ((A \Rightarrow C) \Rightarrow (A \Rightarrow BC))$.

---

* See [3], Chapter 1, Section 10.

III.

III.1. $A \Rightarrow A \vee B$;
III.2. $B \Rightarrow A \vee B$;
III.3. $(A \Rightarrow C) \Rightarrow ((B \Rightarrow C) \Rightarrow (A \vee B \Rightarrow C))$.

IV.

IV.1. $(A \Rightarrow B) \Rightarrow (\bar{B} \Rightarrow \bar{A})$;
IV.2. $A \Rightarrow \bar{\bar{A}}$;
IV.3. $\bar{\bar{A}} \Rightarrow A$.

The axioms are the initial derived formulas. To obtain new derived formulas from those we already have, we take the following two rules of inference:

The substitution rule. If in a derived formula we replace a variable everywhere by an arbitrary formula of the propositional calculus, we obtain another derived formula. We denote by the symbol $\prod_A^\psi (\varphi)$ the formula obtained by replacing the variable A everywhere it appears in the formula φ by the formula ψ.*

Using this notation, we can formulate the substitution rule as follows:

If φ is a derived formula, then $\prod_A^\psi (\varphi)$ is also a derived formula for an arbitrary variable A and an arbitrary formula ψ.

This rule of inference can be written in the following form:

$$\frac{\varphi}{\prod_A^\psi (\varphi)}.$$

(Over the line is written the derived formula that we start with; under it is written the derived formula obtained from it as a result of application of the given rule of inference.)

The detachment rule (DR) (*modus ponens*). If $\varphi \Rightarrow \psi$ and φ are derived formulas, then ψ is a derived formula. With the aid of a

---

* Strictly speaking, we need to show that what we obtain when we replace a letter A everywhere in a formula φ by a formula ψ is always a formula. This is done, for example, in [8], Chapter 2, Section 2.

We note also that the notation $\prod_A^\psi (\varphi)$ does not by any means assume that A necessarily appears in φ. For example, $\prod_B^A (A \vee C)$ coincides with $A \vee C$.

scheme analogous to the one shown above, this rule may be written

$$\frac{\varphi \Rightarrow \psi, \varphi}{\psi}.$$

D. A *proof* or *derivation* of a formula φ from the axioms is a finite sequence of formulas

$\varphi_1, \varphi_2, \ldots, \varphi_n$

satisfying the following two conditions: (1) Every φ_i either is an axiom or is obtained from the preceding formulas in accordance with the substitution or detachment rule; (2) The last formula φ_n is φ. If at least one such sequence of formulas exists, then φ is a derivable (or provable) formula (theorem). Such a decomposition of a proof into elementary steps, each of which consists in writing some axiom or an application of the substitution or detachment rule to either one or two preceding formulas, leads to a large number of steps, so that the proof of a fairly simple formula may turn out to be quite laborious.

As an example, let us derive the formula $A \vee B \Rightarrow B \vee A$ from the axioms; that is, let us show that this is a derivable formula:

1. $(A \Rightarrow C) \Rightarrow ((B \Rightarrow C) \Rightarrow (A \vee B \Rightarrow C));$ (III.3)
2. $(A \Rightarrow B \vee A) \Rightarrow$
   $((B \Rightarrow B \vee A) \Rightarrow (A \vee B \Rightarrow B \vee A));$ $(\prod_C^{B \vee A}(1))$*
3. $B \Rightarrow A \vee B;$ (III.2)
4. $B \Rightarrow C \vee B;$ $(\prod_A^C(3))$
5. $A \Rightarrow C \vee A;$ $(\prod_B^A(4))$
6. $A \Rightarrow B \vee A;$ $(\prod_C^B(5))$
7. $(B \Rightarrow B \vee A) \Rightarrow (A \vee B \Rightarrow B \vee A);$ (2, 6, DR)
8. $A \Rightarrow A \vee B;$ (III.1)
9. $A \Rightarrow A \vee C;$ $(\prod_B^C(8))$
10. $B \Rightarrow B \vee C;$ $(\prod_A^B(9))$
11. $B \Rightarrow B \vee A;$ $(\prod_C^A(10))$
12. $A \vee B \Rightarrow B \vee A.$ (7, 11, DR)

* When the substitution rule is used, we indicate, as in Step 2, what substitution is meant. In addition, the abbreviation DR stands for Detachment Rule.

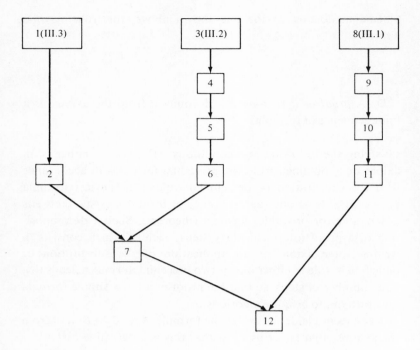

Figure 10

The proof shown on p. 107 in twelve lines can be represented pictorially by the diagram ("tree") in Figure 10.

The proof that we have given can be simplified if, instead of repeated use of the substitution rule (steps 4, 5, 6, 9, 10, and 11), we apply just once a composite substitution rule consisting in the following: If φ is a derivable formula, then

$$\prod\nolimits_{A_1, A_2, \ldots, A_n}^{\varphi_1, \varphi_2, \ldots, \varphi_n} (\varphi)$$

is also a derivable formula, where the latter represents the result of replacing A_1 in φ with the formula φ_1, then replacing A_2 in the formula obtained with the formula φ_2, etc. (Here, we assume that the formulas φ_i do not contain the variables A_i. In this case, the order in which these substitutions are made is immaterial. On the other hand, if the formulas φ_i contain the variables A_i, we can as a preliminary replace these variables in φ with others that do not appear in any of the formulas φ_i.) This composite substitution rule follows immediately from the original substitution rule.*

The symbol

$$\prod_{A_1;\ A_2;\ \ldots;\ A_n}^{\varphi_1;\ \varphi_2;\ \ldots;\ \varphi_n} (\varphi)$$

denotes the same formula as does the symbol

$$\prod_{A_n}^{\varphi_1} (\cdots (\prod_{A_2}^{\varphi_2} (\prod_{A_1}^{\varphi_1} (\varphi))) \cdots).$$

The composite substitution rule can be written in the form of the following scheme:

$$\frac{\varphi}{\prod_{A_1;\ A_2;\ \ldots;\ A_n}^{\varphi_1;\ \varphi_2;\ \ldots;\ \varphi_n}(\varphi)}.$$

Let us now write, and then represent in the form of the "tree" shown in Figure 11, the derivation of the formula $A \lor B \Rightarrow B \lor A$ by applying the composite substitution rule:

1. $(A \Rightarrow C) \Rightarrow ((B \Rightarrow C) \Rightarrow (A \lor B \Rightarrow C));$ (III.3)
2. $(A \Rightarrow B \lor A) \Rightarrow$
   $((B \Rightarrow B \lor A) \Rightarrow (A \lor B \Rightarrow B \lor A));$ $(\prod_{C}^{B \lor A} (1))$
3. $B \Rightarrow A \lor B;$ (III.2)
4. $A \Rightarrow B \lor A;$ $(\prod_{A;\ B;\ C}^{C;\ A;\ B} (3))$
5. $(B \Rightarrow B \lor A) \Rightarrow (A \lor B \Rightarrow B \lor A);$ (2, 4, DR)
6. $A \Rightarrow A \lor B;$ (III.1)
7. $B \Rightarrow B \lor A;$ $(\prod_{B;\ A;\ C}^{C;\ B;\ A} (6))$
8. $A \lor B \Rightarrow B \lor A.$ (5, 7, DR)

INITIAL SUBSTITUTION THEOREM. Any proof of a formula can be transformed into a "proof" of the same formula of the following kind: The only rule of inference used is the detachment rule, and, instead of axioms, we allow the use of any formula obtained from an axiom by a composite substitution (such a formula is called a *substitution instance* of an axiom). To see this, let the given proof $\psi_1, \ldots, \psi_k$ contain k steps and imagine that we know how to transform any proof with fewer than k steps into a proof of the desired kind. There are three cases.

* The class of derivable formulas of the calculus does not change as a result of this.

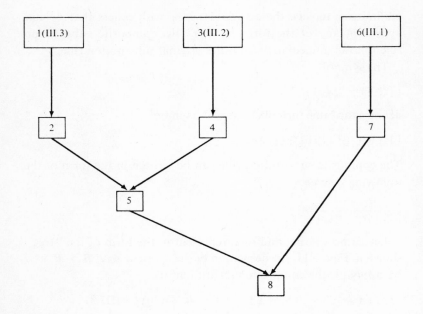

Figure 11

Case 1. ψ_k is an axiom. Then the new proof can be taken to consist of ψ_k alone.

Case 2. ψ_k results from preceding formulas by the detachment rule. Hence, these preceding formulas must be of the form ψ_j and $\psi_j \Rightarrow \psi_k$. By hypothesis, we can construct new proofs of the required kind for ψ_j and $\psi_j \Rightarrow \psi_k$. By combining these proofs and using the detachment rule, we obtain the desired proof of ψ_k.

Case 3. ψ_k results from a preceding formula ψ_j by a substitution; say, ψ_k is $\prod_A^\varphi \psi_j$. By hypothesis, there is a proof of the required kind for ψ_j. In this proof, by replacing all occurrences of A by φ, we obtain a proof of the required kind for ψ_k. (All applications of the detachment rule remain applications of the same rule, and a substitution instance of an axiom becomes a substitution instance of the same axiom.)

Thus, the following alternative approach to the propositional calculus is possible: Take as axioms all substitution instances of the axioms of groups I–IV (pp. 105–106), and use the detachment rule as the sole rule of inference.

The detachment rule also admits a generalization:
If
$\varphi_1, \varphi_2, \ldots, \varphi_n$
and

$$\varphi_1 \Rightarrow (\varphi_2 \Rightarrow (\ldots(\varphi_n \Rightarrow \varphi)\ldots)) \tag{1}$$

are derivable formulas, then φ is also a derivable formula. To see this, note that if Formula 1 and φ_1 are derivable formulas, then, in accordance with the detachment rule,

$$\varphi_2 \Rightarrow (\cdots(\varphi_n \Rightarrow \varphi)\cdots) \tag{2}$$

is also a derivable formula. If Formula 2 and φ_2 are derivable formulas, then, in accordance with the detachment rule,

$$\varphi_3 \Rightarrow (\cdots(\varphi_n \Rightarrow \varphi)\cdots) \tag{3}$$

is also a derivable formula. Continuing our reasoning in this way, we arrive at the formula

$$\varphi_n \Rightarrow \varphi. \tag{n}$$

If formula n and φ_n are derivable formulas, then, in accordance with the detachment rule, φ is also a derivable formula.

Thus, from the detachment rule DR, we have derived a new *composite detachment rule.* (CDR):

$$\frac{\varphi_1 \Rightarrow (\varphi_2 \Rightarrow (\cdots(\varphi_n \Rightarrow \varphi)\cdots)), \varphi_1, \varphi_2, \ldots, \varphi_n}{\varphi}.$$

(A *formal proof of a formula*, defined above in (D), is essentially different from a proof of a rule of inference (e.g. the proof just given of CDR). The first is formulated in the language of the propositional calculus, constituting a finite sequence of derivable *formulas* of this calculus. The second, like the rules of inference themselves, is formulated in a metalanguage.)

The composite detachment rule can be formulated as follows:

If to the axioms we add the formulas $\varphi_1, \varphi_2, \ldots, \varphi_n$ and $\varphi_1 \Rightarrow (\varphi_2 \Rightarrow (\cdots(\varphi_n \Rightarrow \varphi)\cdots))$, then the formula φ is derivable with the aid of the detachment rule (DR) alone.

The derivability of the formula φ with the aid of just the detachment rule (DR) after we have added to the axioms certain formulas $\psi_1, \psi_2, \ldots, \psi_n$ shall mean that there exists at least one finite sequence of formulas satisfying the following two conditions: (1) Every formula in the sequence either is a substitution instance of an axiom, or is one of the ψ_i, or is obtained from preceding formulas in accordance with the detachment rule (DR), and (2) The last formula is φ.

This sequence of formulas is called a **derivation** of the formula φ from the formulas $\psi_1, \psi_2, \ldots, \psi_n$.

If such a sequence of formulas exists, we say that the formula φ is derivable from the formulas $\psi_1, \psi_2, \ldots, \psi_n$, known as the **hypotheses** or **premises**. We write this as follows:

$\psi_1, \psi_2, \ldots, \psi_n \vdash \varphi$.

Thus, the composite detachment rule can be written

$$\varphi_1 \Rightarrow (\varphi_2 \Rightarrow (\cdots(\varphi_n \Rightarrow \varphi)\cdots)), \varphi_1, \varphi_2, \cdots, \varphi_n \vdash \varphi. \tag{a}$$

If the set of premises $\{\psi_1, \psi_2, \cdots, \psi_n\}$ is empty, that is, if $\vdash \varphi$, then φ is simply a derivable (or provable) formula. (This is written: $\vdash \varphi$.) Conversely, if φ is a derivable formula, then, by the initial substitution theorem (cf. pp. 109–110), $\vdash \varphi$. Thus, "$\vdash \varphi$" is equivalent to "φ is derivable."

If $\psi_1, \psi_2, \cdots, \psi_n \vdash \varphi$, one can also easily see that if $\vdash \psi_1$, then $\psi_2, \ldots, \psi_n \vdash \varphi$. If $\vdash \psi_1$ and $\vdash \psi_2$, then $\psi_3, \ldots, \psi_n \vdash \varphi$, etc. If $\vdash \psi_1, \vdash \psi_2, \ldots, \vdash \psi_n$, then $\vdash \varphi$; that is, if all the premises are derivable formulas, then φ is a derivable formula. Thus, we have the following direct consequence of the composite detachment rule (a):

if $\vdash [\varphi_1 \Rightarrow (\varphi_2 \Rightarrow (\ldots(\varphi_n \Rightarrow \varphi)\ldots))]$, then $\varphi_1, \varphi_2, \ldots, \varphi_n \vdash \varphi$.

(aa)

A converse of Proposition aa reads as follows:

if $\varphi_1, \varphi_2, \ldots, \varphi_n \vdash \varphi$, then $\vdash [\varphi_1 \Rightarrow (\varphi_2 \Rightarrow (\ldots(\varphi_n \Rightarrow \varphi)\ldots))]$.

(aaa)

This converse is known as the *deduction theorem*.* It is important since it enables us to establish the derivability of formulas by a shorter procedure.

Let us give an example of the application of the deduction theorem to the derivation of a formula. Let us show that $(A \Rightarrow B) \Rightarrow ((B \Rightarrow C) \Rightarrow (A \Rightarrow C))$ is a derivable formula.

From the formulas $A \Rightarrow B$, $B \Rightarrow C$, and A, we can derive formula C with the aid of the detachment rule alone:

1. $A \Rightarrow B$ (premise)
2. A (premise)
3. B (1, 2, detachment rule)
4. $B \Rightarrow C$ (premise)
5. C (3, 4, detachment rule)

Thus, $A \Rightarrow B$, $B \Rightarrow C$, $A \vdash C$, and, on the basis of the deduction theorem (aaa),

$\vdash (A \Rightarrow B) \Rightarrow ((B \Rightarrow C) \Rightarrow (A \Rightarrow C))$.

As one can see, the proof of the derivability of the formula by using the deduction theorem turned out to be quite simple.

Let us now apply to the formula just derived the following compound substitution: We replace A by the formula φ_1; we replace B by φ_2; and we replace C by φ_3. We obtain

$\vdash (\varphi_1 \Rightarrow \varphi_2) \Rightarrow ((\varphi_2 \Rightarrow \varphi_3) \Rightarrow (\varphi_1 \Rightarrow \varphi_3))$.   (1)

If, in addition,

$\vdash \varphi_1 \Rightarrow \varphi_2$ (2)

and

$\vdash \varphi_2 \Rightarrow \varphi_3$,   (3)

---

* Of course, this is not a theorem of the propositional calculus itself, but rather a theorem about the calculus which is derived and proven contensively. In other words, it is a metatheorem. For a proof of the deduction theorem, see Appendix II (pp. 191–193).

then, from formulas 1, 2, and 3, in accordance with the composite detachment rule, we obtain

$\vdash \varphi_1 \Rightarrow \varphi_3$.

Thus, in the propositional calculus, we can use the rule of inference which may be written as follows:

$$\frac{\varphi_1 \Rightarrow \varphi_2,\ \varphi_2 \Rightarrow \varphi_3}{\varphi_1 \Rightarrow \varphi_3}.$$

This rule is familiar to us as the syllogism rule. However, in the algebra of propositions, it is usually worded "if $\varphi_1 \Rightarrow \varphi_2$ and $\varphi_2 \Rightarrow \varphi_3$ are true, then $\varphi_1 \Rightarrow \varphi_3$ is also true"; in the propositional calculus, it is worded "if $\varphi_1 \Rightarrow \varphi_2$ and $\varphi_2 \Rightarrow \varphi_3$ are derivable, then $\varphi_1 \Rightarrow \varphi_3$ is also derivable."*

EXERCISES

2.39. Show that

$\vdash AB \Rightarrow BA$.

Hint. Apply the composite substitution rule $\prod_{A,C}^{AB,A}$ (Axiom II.3).

2.40. Prove that

$\vdash A \vee A \Rightarrow A$.

Hint: Use the composite substitution rule $\prod_{B,C}^{A,A}$ (Axiom III.3).

2.41. Starting with Hilbert's system of axioms for the propositional calculus [1]:

a1. $A \vee A \Rightarrow A$;
a2. $A \Rightarrow A \vee B$;
a3. $A \vee B \Rightarrow B \vee A$;
a4. $(A \Rightarrow B) \Rightarrow ((C \Rightarrow A) \Rightarrow (C \Rightarrow B))$;

show that $\vdash A \Rightarrow A$.

Hint: Apply the composite substitution rule $\prod_{A,B,C}^{A \vee A, A, A}$ (a4).

2.42. Show that

$A \Rightarrow (B \Rightarrow C), B, A \vdash C,$ \hfill (a)

* The relationship between the truth (identical truth) of formulas in the algebra of propositions and the derivability of formulas in the propositional calculus will be made clear in the following section.

that is, that we can obtain Formula C from the formulas $A \Rightarrow (B \Rightarrow C)$, A, and B with the aid of the detachment rule alone. Apply to (a) the deduction theorem. From the resulting formula, derive the rule of inference:

$$\frac{\varphi_1 \Rightarrow (\varphi_2 \Rightarrow \varphi)}{\varphi_2 \Rightarrow (\varphi_1 \Rightarrow \varphi)}$$

(known as the rule for the permutation of premises).

2.43. Show that

$$\vdash (A \Rightarrow (B \Rightarrow C)) \Rightarrow (AB \Rightarrow C).$$

Show that from this formula we can obtain the rule of inference

$$\frac{\varphi_1 \Rightarrow (\varphi_2 \Rightarrow \varphi)}{\varphi_1 \varphi_2 \Rightarrow \varphi}$$

(known as the rule for the combination of premises).

3. CONSISTENCY, INDEPENDENCE, AND COMPLETENESS OF A SYSTEM OF AXIOMS IN THE PROPOSITIONAL CALCULUS

3.1. The consistency problem is the most fundamental of the problems arising in the construction of a calculus (formalized language, axiomatic system). There are two conceptions of consistency corresponding to two aspects of the study of a formalized language, namely, the syntactical and the semantic. Syntax studies the elements and the structure of the formalized language without regard to what it expresses. Semantics studies the elements and structure of a formalized language in connection with its meaningful interpretation (in connection with what it expresses of extralinguistic reality).

From the syntactical point of view, a system of axioms (containing a sign for negation*) is said to be **consistent** if there is no way of deriving from it two formulas one of which is the negation of the other, that is, no way of deriving two formulas of the forms φ and $\bar{\varphi}$. If the condition of consistency is not satisfied, that is, if there exists at least one pair of derivable formulas of the forms φ and $\bar{\varphi}$,

* With its usual properties (let us say, expressed in axioms IV.1–IV.3 [see p. 106]).

then such a (contradictory) system of axioms cannot serve as the basis for constructing a calculus.* Consistency in this syntactical sense is also called freedom from internal contradiction.

From the semantic point of view, a system of axioms is said to be consistent if it has at least one model. In this case, an axiomatic system is also said to be **interpretable** or **realizable**. (Consistency in this semantic sense is also called external freedom from contradiction.)

If a system of axioms is realizable, there exists at least one system of objects (individuals and relations) the structure of which is described by the given axiomatic theory. This means that all the derivable formulas in the theory can be interpreted as true propositions expressing properties of the structure of that system of objects (model). Thus, an arbitrary derivable formula φ becomes a true proposition relating to the model. Then, $\bar{\varphi}$ is a false proposition and hence cannot be a derivable formula; that is, the system of axioms is free of internal contradiction because it is realizable. There is, however, one very important condition: in this reasoning, we have assumed implicitly that the *model itself* of the axiomatic system is free of internal contradiction. The existence of a model of an axiomatic system merely reduces the question of consistency of this system to the question of consistency of the system of concepts of which this model is constructed. If this system is free of internal contradiction, then the given axiomatic system is also free of internal contradiction.

The method of models was widely employed in the second half of the nineteenth century (and is employed today) to reduce the question of consistency of one axiomatic system to the question of consistency of another axiomatic system.

Thus, for example, by constructing a model of the geometric system of Lobachevski in terms of Euclidean geometry, the question of consistency of Lobachevskian geometry can be reduced to the question of consistency of Euclidean geometry. By constructing a model of the latter in terms of the arithmetic of real numbers,

---

* This is true because, in such a system, any formula can be proved (and, at the same time, refuted), so that the concept of a theorem loses all significance.

the question of its consistency can be reduced to the question of the consistency of the arithmetic of real numbers.

The system of real numbers can be constructed as a consistent extension of the system of rational numbers, the system of rational numbers can be constructed as a consistent extension of the system of integers, and the system of integers can be constructed as a consistent extension of the system of natural numbers. The question of the consistency of the arithmetic of real numbers can, in the final analysis, be reduced to the question of the consistency of the arithmetic of natural numbers.

For the arithmetic of natural numbers, one can construct a set-theoretic model, so that the question of its consistency reduces to the question of the consistency of set theory.

Thus, the question of the consistency of all the axiomatic systems enumerated here reduces to that of the consistency of set theory. Since a set-theoretic model can be found for an arbitrary consistent system of axioms, the method of models is no longer applicable to a solution of the question of the consistency of set theory itself, that is, of the system of axioms of that theory (its application would lead to a vicious circle), and the consistency of set theory must be assumed, although serious difficulties of a logical nature do arise in that theory that keep us from being intuitively sure of its consistency.*

Hilbert proposed a new procedure for proving consistency, one that does not involve the method of models. Hilbert's program assumes the existence of ways of describing the class of derivable (true) formulas of a given calculus. In this case, it may be possible to obtain a direct proof of the impossibility of the existence, in this calculus, of derivations of two formulas of which one is the negation of the other, that is, a direct proof of freedom from internal contradiction.

Let us prove the consistency of the axiomatic system of the propositional calculus that was described in Section 2. We set ourselves the problem of establishing that, if φ is a formula that is

* See [3], Introduction, and [5], Chapter 3, Part 1.

derivable in the propositional calculus, then $\bar{\varphi}$ is not derivable. To solve this problem, we assign to the symbols of the alphabet of the propositional calculus the same meaningful interpretation as in the algebra of propositions. With this interpretation of the symbols of the alphabet, all formulas of the propositional calculus become formulas of the algebra of propositions and all derivable formulas of the propositional calculus become tautologies in the algebra of propositions.

The first of these statements is obvious; the second is established as follows: First of all, let us show that all the axioms of the propositional calculus are tautologies in the algebra of propositions. This is easily proved, for example, with the aid of truth tables. (The reader can do this himself without difficulty.*)

It remains to show that application of the rules of inference (the substitution rule and the detachment rule) to tautologies again leads to tautologies. To see this, let φ denote a tautology containing the variable A. Let us replace A everywhere in the formula φ by an arbitrary formula ψ. Since the formula ψ, like A, assumes only the values T and F, the formula $\prod_{A}^{\psi}(\varphi)$ obtained as a result of this substitution is also a tautology; that is, $\prod$ preserves tautologicality.

Let us show that the detachment rule also preserves tautologicality. If φ and $\varphi \Rightarrow \psi$ are tautologies, then ψ is also a tautology because, if ψ assumed the value F for some combination of values of the variables, then $\varphi \Rightarrow \psi$ would become T $\Rightarrow$ F; that is, it too would assume the value F, which contradicts the assumption. Thus, since the axioms are tautologies and application of the rules of inference to tautologies leads to tautologies, all the derivable formulas of the propositional calculus, considered as formulas in the algebra of propositions, are tautologies. It follows immediately from this that the propositional calculus is consistent. To see this, let φ denote an arbitrary formula that is derivable in the propositional calculus. Then, φ is a tautology in the algebra of propositions and $\bar{\varphi}$ is an identically false formula. Since $\bar{\varphi}$ is not a tautology in the algebra of propositions, it cannot be derived in the propo-

---

* See Exercise 1.22.

sitional calculus. We have shown that there does not exist a pair of formulas of the form φ and $\bar{\varphi}$ of derivable formulas; that is, the propositional calculus is consistent.

3.2. An axiom (belonging to some system of axioms) is said to be **independent** if it cannot be derived from the remaining axioms of the system. A system all the axioms of which are independent is called an **independent system**. Such a system contains no "superfluous" axioms that can be derived from the other axioms.

The system of axioms of the propositional calculus given in Section 2 is independent. Proof of the independence of a system of axioms consists in proving the independence of each axiom of the system from the system defined by the other axioms.

To prove the independence of any axiom A_k, we need only construct an interpretation of the symbols of the alphabet in such a way that all the axioms of the given system except A_k and all formulas that can be derived from these axioms assume the value T for all combinations of values of variables and that the axiom A_k will assume the value F for some combination of values of the variables. In such a case, the axiom A_k is independent because, if it were derivable from the other axioms, it would assume the value T for all combinations of values of the variables.

As an example, let us prove the independence of Axiom IV.1:

$(A \Rightarrow B) \Rightarrow (\bar{B} \Rightarrow \bar{A})$.

Suppose that the variables of the calculus assume the two values T and F, that is, the values of the variables of the algebra of propositions.

We define implication, conjunction, and disjunction in the same way as in the algebra of propositions. We define the operation of negation as follows:

A	$\bar{A}$
T	T
F	F

that is, $\bar{A}$ equiv A.

Under this interpretation, the axioms of the first three groups assume, for arbitrary sets of values of the variables, the value T since they do not contain the negation symbol, and the remaining operations are defined the same way as in the algebra of propositions, where these axioms are tautologies.

One can easily see that axioms IV.2 and IV.3 also assume the value T for arbitrary sets of values of the variables:

A	$\bar{A}$	$\bar{\bar{A}}$	$A \Rightarrow \bar{A}$ (IV.2)	$\bar{\bar{A}} \Rightarrow A$ (IV.3)
T	T	T	T	T
F	F	F	T	T

The formulas derived from axioms I–III, IV.2, and IV.3 also assume the value T for arbitrary combinations of values of the variables since the application of the rules of inference to tautologies yields tautologies (as was shown above in the proof of the consistency of the propositional calculus).

Let us show that Axiom IV.1 does not assume the value T for all combinations of values of the variables:

A	B	$\bar{A}$	$\bar{B}$	$A \Rightarrow B$	$\bar{B} \Rightarrow \bar{A}$	$(A \Rightarrow B) \Rightarrow (\bar{B} \Rightarrow \bar{A})$
T	T	T	T	T	T	T
T	F	T	F	F	T	T
F	T	F	T	T	F	F
F	F	F	F	T	T	T

Thus, if A assumes the value F and B assumes the value T, then Axiom IV.1 assumes the value F. This proves the independence of Axiom IV.1. To prove the independence of each of the axioms, a special interpretation is made. Consequently, to prove the independence of the system of axioms I–IV, we need to construct eleven interpretations.*

Probably the best known of the consistency problems is the problem of Euclid's fifth postulate. Throughout the course of

---

* See [8], Chapter 2, Section 11, p. 79.

two millennia, up to the second half of the nineteenth century, many mathematicians assumed that the Euclidean axiom regarding parallel lines* could be derived from the other geometrical axioms and they sought such a derivation. The errors made, even by famous mathematicians, in the attempts to solve this problem are explained by the state of science of their time. In particular, they did not yet have a complete list of geometrical axioms, and the logical tools for deduction were not precise. We know of numerous "proofs" of the fifth postulate but they are not valid because, in each of them, the axioms used include (implicitly and unnoticed by the author of the proof) an assumption equivalent to the fifth postulate relative to the other axioms, that is, a proposition that itself could be derived from these axioms if we combine the fifth postulate with them.

If we denote by G the set of geometrical axioms excluding the postulate of parallels (here, G is the system of axioms of *absolute* geometry) and if we denote by V the fifth postulate, then the problem of the fifth postulate can be formulated as follows: Prove that G implies V.

In the "proofs" of the fifth postulate referred to above, what is usually shown is that

G, A imply V,

where A is a proposition such that the proposition

"G, V imply A"

is true but the proposition

"G implies A"

is not true.

Thus, what each of these "proofs" does essentially is not to solve the problem of the fifth postulate but to provide some equivalent A of this postulate relative to the remaining axioms, that is, a proposition from which V can be derived and which can itself be derived from V on the basis of the system of axioms G

---

* The fifth postulate in Euclid's *Elements*.

(using a logical system as the logical language of geometrical theory). Some examples of such equivalents of the fifth postulate are: The distance between points on one straight line and points on a line parallel to it is bounded below by a positive number (Proclus' proof). Similar triangles exist (Wallis' proof). Through every point inside an angle, one can draw a straight line intersecting both sides of the angle (Legendre's proof).

As we know, the problem of the fifth postulate was solved in a negative way: It was shown that the Euclidean axiom regarding parallels is *independent* of the remaining axioms in Euclid's geometry. The system of axioms in Lobachevski's geometry differs from the system of axioms in Euclid's geometry only as regards the single axiom on parallels. We recall that Euclid's fifth postulate V is as follows: for an arbitrary straight line and a point not on that straight line, there exists in the plane defined by that line and that point exactly one straight line passing through the given point parallel to the given line (that is, not intersecting it). Lobachevski's axiom V' on parallels is a negation of V: for an arbitrary straight line and a point not on it, there exist in the plane defined by that line and that point more than one straight line passing through the given point and parallel to the given line. Thus, G, V is the system of axioms of Euclidean geometry and G, V' is the system of axioms of Lobachevskian geometry.

Therefore, every model of the system of axioms G, V' of Lobachevskian geometry is a system of objects in which all the axioms belonging to G are satisfied (are true propositions) and V is not satisfied (becomes a false proposition), whereas its negation V' is satisfied (is true). The existence of at least one such model (the interpretation of Lobachevskian geometry in terms of Euclidean geometry) proves the independence of V.

The idea of such a proof of the consistency of Lobachevskian geometry relative to Euclidean geometry was first proposed by the Italian mathematician E. Beltrami (1868). The final solution of this problem was obtained by the German mathematician F. Klein and the French mathematician H. Poincaré.

3.3. The completeness of a system of axioms is understood in two senses. A system of axioms constituting the basis of some calculus is said to be **complete in the broad sense** if every meaningfully true proposition in the domain of objects described by this calculus can be derived in the calculus. The system of axioms of the propositional calculus given in Section 2 is complete in this sense; that is, every tautology of the algebra of propositions is a derivable formula of the propositional calculus.* Furthermore, since in our proof of the consistency of the propositional calculus we showed that every formula that can be derived in this calculus is, when considered as a formula in the algebra of propositions, a tautology, it follows that the class of derivable formulas of the propositional calculus coincides with the class of tautologies of the algebra of propositions.

Therefore, to answer the question as to the derivability of some formula in the propositional calculus, we do not need to find its derivation from the axioms. We need only show that this formula, regarded as a formula in the algebra of propositions, is a tautology. From this it follows that the decision problem of the propositional calculus, that is, the question of existence of a finite procedure enabling us to determine for a specific formula in this calculus whether it is derivable or not, is answered in the affirmative.†

The concept of completeness in the broad sense is connected with a meaningful interpretation of the system of axioms, that is, it has a semantic character. We can speak of such completeness under the condition that the system of axioms has an interpretation, that is, is semantically consistent.

However, the consistency of a system of axioms can be treated also in its syntactical aspect, as internal consistency. Corresponding to this concept of consistency there is a concept of completeness in another, more restricted, sense: A system of axioms is said to be **complete in the narrow sense** if any extended system of axioms obtained by adding to the axioms of the given system some formula not derivable from them is inconsistent. (One can easily

---

* For a proof, see Appendix III (pp. 193–197).
† In Chapter 1, Sections 2 and 4, we exhibited two decision procedures, namely, the construction of a truth table and the reduction of a formula to conjunctive normal form.

see that completeness of a system in the narrow sense implies completeness in the broad sense since a formula the addition of which to the system makes it contradictory cannot be true in any interpretation.)

The system of axioms of the propositional calculus that was given in Section 2 is also complete in this sense. Let us prove this.

Let φ denote any formula in conjunctive normal form that is not a tautology of the algebra of propositions. Then, φ cannot be derived in the propositional calculus (since every formula that is derivable in the calculus of propositions is a tautology in the algebra of propositions).

Let us add the formula φ to the axioms of the propositional calculus. In this extended system of axioms, the formula φ itself and each of its conjuncts are derivable formulas.

Since φ is not a tautology, there exists at least one conjunct of the conjunctive normal form that is an elementary disjunction which contains no variable together with its negation, that is, which is of the form

$$X_1^{\alpha_1} \vee X_2^{\alpha_2} \vee \cdots \vee X_k^{\alpha_k}$$

where each elementary variable occurs only once (either with or without the symbol for negation).

In this disjunction, let us make the following substitution:

$\prod_{x_1^{\alpha_1}, x_2^{\alpha_2}, \ldots, x_k^{\alpha_k}}^{A, A, \ldots, A}$. This yields $A \vee A \vee \cdots \vee A$,

or, in accordance with the idempotency law, A; that is, A is a derivable formula. But we can replace A with $\bar{A}$ and obtain the result that $\bar{A}$ is also a derivable formula. Thus, we have obtained the result that the extended system of axioms is contradictory. Consequently, the given system of axioms (Section 2) is complete in the narrow sense.

EXERCISES

2.44. Show that the axioms I–IV of the propositional calculus are tautologies of the algebra of propositions.

2.45. Consider the following interpretation for the symbols of the alphabet of the propositional calculus:
a. The variables assume the two values T and F.
b. Implication, disjunction, and negation are defined the same way as in the algebra of propositions.
c. Conjunction is defined as follows: $AB =_{Df} B$,*
that is,

A	B	AB
T	T	T
T	F	F
F	T	T
F	F	F

1. Determine what values axioms I–III assume for all possible combinations of the values of the variables.
2. Is the independence of any axiom proved with the help of the given interpretation?

2.46. Do the same thing for an interpretation of the symbols that coincides with that of the preceding problem as regards (a) and (b) but in which conjunction is defined by $AB =_{Df} F$.

---

* The symbols $=_{Df}$ and, later on, equiv$_{Df}$ mean "equivalent (or equal) by definition" (from the Latin *definitio*). These are metalinguistic symbols.

3
PREDICATE LOGIC

1. SETS. OPERATIONS ON SETS

In what follows, we shall use certain concepts from the algebra of sets. The present section is devoted to clarifying these concepts and making them precise.*

1.1. The concept of a set is not defined in terms of other concepts. Intuitively, we understand by a set a class, collection, aggregate, etc., of any objects distinct from each other. What their nature is is completely immaterial (The words "set," "class," "collection," and "aggregate" are all synonyms.)

An object belonging to a set of objects is called an **element** of that set. The proposition "the object a belongs to the set S" or, in slightly different wording, "the object a is an element of the set S" is denoted in symbols by writing $a \in S$.

A set can be considered given or known if we have some means of determining, for an arbitrary given object, whether it does or does not belong to the set. A finite set can be given by a direct listing or enumeration of all its elements (in any order). In this case, the custom is to list the objects one after another and enclose the list in braces. For example, the set of digits used in the decimal numeration system is written

$\{0, 1, 2, 3, 4, 5, 6, 7, 8, 9\}$.

When a set is defined in this way, we can determine, for any object, whether it does or does not belong to the set simply by comparing it with the listed elements of that set. If it coincides with any element of the set, it belongs to the set; otherwise, it does not belong to it. (Of course, it is assumed that we know how to identify the object and to distinguish it from any other object.)

An infinite set cannot be defined by listing its objects.

A set, finite or infinite, can be defined by a characteristic property, that is, a property possessed by every element of the set but not possessed by any object that is not an element of the set. We denote the set characterized by the property P by†

---

* We recall that our treatment of the concept of a function (see Introduction, p. 6) was made on a set-theoretic basis.
† This set is also denoted by the symbol $\hat{x}[P]$.

$\{x : P\}$

(read "the set of all x possessing property P").

For example, $\{x : x > 0\}$ is the set of positive numbers, that is, the set of all those numbers the substitution of which for x in $x > 0$ yields a true proposition (for example, $2 > 0$, $3.5 > 0$, etc.).

When a set is defined in this way, the possibility of determining whether or not an arbitrary given object belongs to the set in question by verifying whether or not the object possesses the property P or not depends, of course, on the property P itself.*

If two sets A and B consist of exactly the same elements, we say that they are equal and we indicate this fact by the usual equality sign:

$A = B$.

Sets A and B can be defined by means of different properties P and Q but if possession of property P on the part of an object implies its possession of property Q and vice versa, then these sets are equal.

Suppose that A is the set of triangles two of whose sides are equal and that B is the set of triangles two of whose angles are equal. Since the existence of two equal sides of a triangle implies the existence of two equal angles and vice versa, these sets are equal; that is, $A = B$.

1.2. Let A denote a given set. Every set B consisting of elements of the set A is called a **subset** of A. We also say that B is **included in** A.

Here, if there exist elements of the set A that do not belong to B, we say that B is **properly included** in A or that B is a **proper subset** of A. We denote this fact by writing $B \subset A$. If we wish to indicate that B is included in A but do not wish to restrict ourselves to proper

---

* A simple example: The set consisting of a *single* natural number $n > 2$ that is the least of all natural numbers $n > 2$ for which there exist positive integers x, y, and z such that $x^n + y^n = z^n$. One can easily see that finding such an n would constitute a refutation of the famous theorem of Fermat and a proof that this set is empty would constitute a proof of that theorem.

inclusion, that is, if we wish to allow the possibility that $B = A$, we write $B \subseteq A$.

We denote by lower-case letters near the end of the alphabet (x, y, z) with or without subscript, variables for elements of a set, that is, variables in place of which one can substitute the names of arbitrary elements of that set. These variables are called **individual variables**. The elements of a set D with whose names we decide to replace the variable are called the **values** of that variable and the set D is called the **range of values** of the variable. For names of definite (fixed) elements of a set, we use lower-case letters near the beginning of the alphabet (a, b, c), again with or without subscripts. We call these elements **individual constants**.

The notation $B \subseteq A$ (read "the set B is included in the set A") expresses the same thing as the statement

for all x, $(x \in B) \Rightarrow (x \in A)$

(read "for all x, if x belongs to the set B, then it also belongs to the set A"). The relationship of inclusion (not restricted to proper inclusion) is reflexive ($A \subseteq A$), antisymmetric ($B \subseteq A$) $\wedge$ ($A \subseteq B$) $\Rightarrow$ ($A = B$), and transitive ($A \subseteq B$) $\wedge$ ($B \subseteq C$) $\Rightarrow$ ($A \subseteq C$).

The relation of proper inclusion is antireflexive: $\overline{A \subset A}$.

A subset B of a set A can be defined by means of a property P:

$B = \{x : x \in A \wedge P\}$.

(read "B is the set of all x that belong to the set A and possess property P").

For example, if A is the set of parallelograms and P is the property of having a right angle, then B is the set of all parallelograms that have a right angle (that is, it is the set of all rectangles).

If all elements of the set A possess property P, then $B = A$. The set A is considered a subset of itself (we sometimes say that it is an *improper* subset of itself).

For example, if A is the set of parallelograms and P is the property

"the diagonals bisect each other,"

then $B = A$, since the diagonals of every parallelogram bisect each other.

If a set A does not have a single element possessing property P, then the set B does not have any element at all.

A set with no elements at all is called the **empty set** and is denoted by the symbol $\varnothing$. The empty set is also considered a subset of an arbitrary set A (and we sometimes speak of it as the empty subset).

For example, if A is the set of parallelograms and P is the property of having five sides, then $B = \varnothing$ since a parallelogram with five sides does not exist.

Suppose that the set A is the set $\{a, b, c\}$. Then the subsets of A are the following sets:

$\varnothing$, $\{a\}$, $\{b\}$, $\{c\}$, $\{a, b\}$, $\{a, c\}$, $\{b, c\}$, $\{a, b, c\}$.

1.3. Suppose that $A, B, C, \ldots$ are subsets of some set, which we shall call the **universal set** and shall denote by U.

We shall define three operations on the set of all subsets of the universal set (including $\varnothing$ and U itself): complement, intersection, and union.

These operations on sets generate new sets. Just as in the algebra of propositions, we use the same terminology in the case of each of these operations for the operation itself and the result of applying it.

In the following definitions of these operations, we shall be referring to the results of application of the operation in question.

We use the individual variables x, y, z with range of values U.

a. The **complement** of a set A is defined as the set consisting exclusively of all elements of the universal set U that do not belong to A.

We denote the complement of the set A by $\overline{A}$.

The definition of the complement of a set can be written as follows:

$\overline{A} = \{x : \overline{x \in A}\}$.

$\overline{A}$ is, by definition, the set of all x (in U) that do not belong to A.

b. The **intersection** of sets A and B is defined as the set consisting exclusively of all those elements that belong both to the set A and to the set B.

The intersection of the sets A and B is denoted by $A \cap B$. Thus,

$$A \cap B = \{x : x \in A \land x \in B\}$$

c. The **union** of sets A and B is defined as the set consisting exclusively of all elements that belong to at least one of the sets A or B.

The union of the sets A and B is denoted by $A \cup B$. Thus,

$$A \cup B = \{x : x \in A \lor x \in B\}$$

The algebra of sets, which describes the structure of the set of subsets of some set (the universal set) together with these three operations, is, like the algebra of propositions, a model of the abstract Boolean algebra described in Section 4 of Chapter 1 (p. 73). All properties of the operations of the algebra of propositions (negation, conjunction, and disjunction) can be carried over to the language of the algebra of sets by means of the following dictionary:

Language of the Algebra of Propositions		Language of the Algebra of Sets	
$A, B, C, \ldots$	Variables for propositions	$A, B, C, \ldots$	Variables for sets (subsets of some universal set)
—	Negation	—	Complement
$\land$	Conjunction	$\cap$	Intersection
$\lor$	Disjunction	$\cup$	Union
T	True proposition	U	Universal set
F	False proposition	$\emptyset$	Empty set
Equiv	Equivalence of formulas	$=$	Equality of sets

For example, the law of double negation (1) $\bar{\bar{A}}$ equiv A (or $\vdash \bar{\bar{A}} \Leftrightarrow A$) becomes the law for the complement of the complement: $\bar{\bar{A}} = A$.

This equation can be established by beginning with the definition of the complement and using the law of double negation:*

$$\bar{\bar{A}} = \{x : \overline{x \in \bar{A}}\} = \{x : \overline{\overline{x \in A}}\} = \{x : x \in A\} = A.$$

Analogously, de Morgan's law (17) $\overline{A \vee B}$ equiv $\bar{A} \wedge \bar{B}$ (or $\vdash \overline{A \vee B} \Leftrightarrow \bar{A} \wedge \bar{B}$) is translated into the language of the algebra of sets in the form of the equation $\overline{A \cup B} = \bar{A} \cap \bar{B}$, as we easily show:

$$\overline{A \cup B} = \{x : \overline{(x \in A) \vee (x \in B)}\} = \{x : \overline{x \in A} \wedge \overline{x \in B}\}$$
$$= \{x : \overline{x \in A}\} \cap \{x : \overline{x \in B}\} = \bar{A} \cap \bar{B}.$$

The properties of the operations of the algebra of sets are shown pictorially by means of a geometrical representation. Figure 12 represents the universal set U in the form of an arbitrary rectangle and its subsets in the form of figures (in the present case, we chose circles) lying inside that rectangle. Figure 12a represents the set $\overline{A \cap B}$ by diagonal shading; Figure 12b represents the set $\bar{A}$ by horizontal shading and the set $\bar{B}$ by vertical shading. The

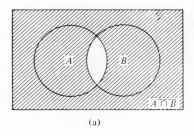
(a)

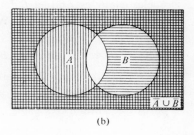
(b)

Figure 12

* Here, we treat the expression $x \in A$ as a variable for a proposition that assumes the values T and F disregarding the fact that this variable is, in turn, a function of the individual variable x (we shall speak further about this later) and we apply to it the operations of the algebra of propositions.

portion of this rectangle with shading of any sort represents the set $\bar{A} \cup \bar{B}$. As one can see, the sets $\overline{A \cap B}$ and $\bar{A} \cup \bar{B}$ consist of the same points of the rectangle; that is,

$$\overline{A \cap B} = \bar{A} \cup \bar{B}.$$

This equation is a translation of de Morgan's law (16) $\overline{A \wedge B}$ equiv $\bar{A} \vee \bar{B}$ or $\vdash \overline{A \wedge B} \Leftrightarrow \bar{A} \vee \bar{B}$ into the language of the algebra of sets.

EXERCISES

3.47. Translate into the language of the algebra of sets properties 1–17 of operations in the algebra of propositions. Prove the resulting properties by using the definitions of the operations of the algebra of sets and the analogous properties of the operations of the algebra of propositions. Verify these properties by using a geometrical representation.

3.48. Suppose that $A \subset B$. To what are $A \cup B$ and $A \cap B$ equal? What is the relationship between the complements $\bar{A}$ and $\bar{B}$?

3.49. What is the number of subsets of a set consisting of two elements; of three elements; of n elements? Among the subsets of a set of n elements, how many are there that contain one element? two elements?...$(n - 1)$ elements?

3.50. Let S denote the set of triangles, let A denote the set of isosceles triangles, and let B denote the set of right triangles. Treat the set S as the universal set. What, in everyday language, are the sets $A \cup B$, $A \cap B$, and the complements (with respect to S) of the sets $A, B, A \cup B$, and $A \cap B$?

3.51. We say that a set S is partitioned into two classes A_1 and A_2 if $A_1 \cup A_2 = S$ and $A_1 \cap A_2 = \emptyset$. More generally, we say that S is partitioned into n classes $A_1, A_2, \ldots, A_n$ if

$$\bigcup_{i=1}^{n} A_i = A_1 \cup A_2 \cup \ldots \cup A_n = S$$

and

$$A_i \cap A_j = \emptyset$$

whenever $i \neq j$ and $i, j = 1, 2, \ldots, n$. (The definitions of union and intersection, like the definitions of disjunction and conjunction, can be extended to an arbitrary number of sets.)

Each set A partitions the universal set into two classes (subsets) A and $\bar{A}$:

$$A \cup \bar{A} = U \wedge A \cap \bar{A} = \emptyset.$$

Show that, in general, two sets A and B partition the universal set into four classes. (Of course, some of these classes may be empty.)

Hint: Write these classes with the aid of the sets A, B, $\bar{A}$, and $\bar{B}$ and the operations $\cup$ and $\cap$. Then verify that the defining conditions for a partition of a set into classes are satisfied.

3.52. Show that, in general, three subsets A, B, and C of the universal set partition it into eight classes.

3.53. We say that two sets whose intersection is different from the empty set **overlap**. Two sets possessing no common elements, that is, whose intersection is empty, are said to be **disjoint**. Indicate which of the following pairs of sets are overlapping and which are disjoint and, in the former case, write their intersection.

    a. The set of rational numbers and the set of positive numbers;
    b. the set of equilateral triangles and the set of right triangles;
    c. the set of pyramids and the set of regular polyhedra;
    d. $\{x = x \in N \wedge x < 5\}$ and $\{x : x \in N \wedge x > 3\}$
(where N is the set of natural numbers);
    e. $\{x : x \in C \wedge x \leq 4\}$ and $\{x : x \in C \wedge x > 4\}$
(where C is the set of integers);
    f. $\{x : x \in C \wedge x < 4\}$ and $\{x : x \in C \wedge x > 4\}$.

3.54. Let us denote by $m(A)$ the number of elements of a set A. Prove the following relations:

$$m(A \cup B) = m(A) + m(B) - m(A \cap B);$$
$$m(A \cup B \cup C) = m(A) + m(B) + m(C) - m(A \cap B) - m(B \cap C)$$
$$- m(A \cap C) + m(A \cap B \cap C).$$

Write these relations for the case in which A, B, and C are pairwise disjoint.

2. THE INADEQUACY OF PROPOSITIONAL LOGIC. PREDICATES

2.1. Using propositional logic, we have made an analysis of certain lines of reasoning (see Chapter 1, Section 3) to show their applicability. We noted that, if, in the derivation of certain propositions from other propositions, we consider the internal structure of the elementary propositions, then to show the applicability of such reasoning, the tools of propositional logic prove insufficient.

For example, the legitimacy of the derivation

$$\frac{\text{every integer is a rational number,}}{\text{consequently, 1 is a rational number}} \tag{1}$$

cannot be established with the tools of propositional logic since in this derivation the premises and the conclusion are, from the point of view of this logic, elementary propositions treated as units, indivisible, without regard to their internal structure.

However, even this derivation is formal and does not depend on the content of the premises and conclusion but only on their form, their structure.

One can easily see that the reasoning

$$\frac{\text{every rhombus is a parallelogram,}}{\text{consequently, } ABCD \text{ is a rhombus;}} \tag{2}$$

although it differs in content from 1, has exactly the same structure, and the drawing of the conclusion from the premises in these two cases is determined by their structure and not their content.

In the language of propositional logic, we cannot separate the logical structure of these two lines of reasoning from their particular content. If we tried to do this, replacing each elementary proposition with a propositional variable, both lines of reasoning would take the form

X and Y imply Z.

However, it is perfectly obvious that such a representation of these two lines of reasoning does not enable us to see whether Z actually does follow from the premises X and Y or not, since the structure of the premises and the conclusion is not reflected in this representation.

Thus, propositional logic does not provide us with the tools for a sufficiently fine analysis of these lines of reasoning to tell whether they are legitimate or not. This is explained by the fact that propositional logic is limited to reducing compound propositions to elementary ones. It regards compound propositions as functions of elementary ones but it does not decompose elementary propositions, further although these are not the simplest elements in our reasoning and they possess an internal structure that plays an important role in deduction.

Therefore, in the language of propositional logic, we cannot obtain all those properties of a derivation that are necessary for the axiomatic construction of various mathematical theories.

It becomes necessary to broaden propositional logic and to construct a logical system by means of which we can also investigate the structure of the elementary propositions. Such a logical system is *predicate logic*, which contains propositional logic as one of its branches.

2.2. In traditional logic, we single out a subject (the thing about which one makes an assertion in the proposition) and a predicate (the thing that is asserted about the subject) in an elementary proposition.

In the proposition

A rhombus is a parallelogram,

"rhombus" is the subject and "parallelogram" is the predicate (the word "is" is the copula). This proposition can be interpreted as the assertion that a rhombus possesses the property of "being a parallelogram" or that the set of rhombuses is contained in the set of parallelograms.

This division into a subject and predicate, which is characteristic of traditional logic, is possible (and sufficient) only in those cases in which the elementary proposition expresses a property of an object but is not suitable when the elementary proposition expresses a relationship between objects.

For example, the elementary propositions

The number 2 is less than the number 3,
The point A lies between the points B and C, etc.

cannot be represented in the form "S is P," where S is the subject and P is the predicate.*

Predicate logic also begins with the division of elementary propositions into a subject (or subjects) and a predicate, but this division is not done in the same way as in traditional logic.

Here, the predicate is treated as a logical function of one or several individual variables (depending on whether the proposition expresses a property of an object or a relationship between two or more objects). Such a representation is suitable both for those elementary propositions that express the properties of objects and for those that express relationships between objects.

Predicates are denoted symbolically by functional signs with one or more blank spaces or with one or several variables occupying the blank spaces.

Let us look at some examples.

a. The proposition

5 is a prime number

is a true proposition; the proposition

4 is a prime number

is a false proposition; the expression

... is a prime number

---

* Of course, we can, for example, consider A in the second of these examples as the subject and the property "lying between B and C" as the predicate (and we shall encounter representations of this kind later); but, in the first place, such a representation is not unique (for example, we can think of the proposition as being a proposition regarding either B or C as well as A) and, in the second place, there arises a not always desirable "coalescence" of the different subjects (in the present case, B and C) in a single predicate.

is not a proposition at all since we cannot say either that it is true or that it is false. This is a logical function that becomes a true or a false proposition when the dots are replaced by a natural number. This logical function is also called a **one-place** predicate. We denote it by $P(\ )$, using the functional sign P with a blank space, or by $P(x)$ since, instead of "...is a prime number," we can say "x is a prime number," where x is a variable for numbers belonging to some specified set, in the present case, the set of natural numbers.

Adopting this convention, we denote by $P(5)$ the true proposition "the number 5 is a prime number" and we denote by $P(4)$ the false proposition "the number 4 is a prime number." We denote by $P(x)$ the predicate or logical function "x is a prime number," this predicate becoming a true or a false proposition depending on the value substituted for x.

The domain of definition of this logical function is the set N of natural numbers and its range of values is the set $\{T, F\}$; that is,

$$N \xrightarrow{P} \{T, F\}.$$

The predicate $P(x)$ partitions the domain of definition into two subsets on one of which this predicate becomes a true proposition (that is, every number in this subset converts it into a true proposition) and on the other, a false proposition. Of these two subsets of the domain of definition of the predicate $P(x)$, we shall refer to the one that converts this predicate into a true proposition as the **truth set** of that predicate.

The notation $\{x : P\}$ (which we used above in Section 1) for the set of objects possessing property P is the notation for the truth set of the predicate $P(x)$. If, let us say, $P(x)$ is the notation for the predicate "x is a prime," then $\{x : P(x)\}$ denotes the set of prime numbers.

We can understand by $P(x)$ an *arbitrary* one-place predicate. In this case, P plays the role of a variable (known as a **predicate variable**) the values of which are various specific one-place predicates.* In our present example, we had only one of the value of this predicate variable.

Let us look at another example. This time, let $P(x)$ denote the predicate "the city x is the capital of France." Then, the expression P(Paris) denotes the true proposition "the city Paris is the capital of France," and the expression P(London) denotes the false proposition "the city of London is the capital of France." In this case, the domain of definition of the predicate $P(x)$ is the set of cities; that is, in place of the predicate variable x, we decide to substitute the name of an arbitrary city, and the truth set consists of the one city the substitution of whose name for the variable x converts this predicate into a true proposition:

$\{x : P(x)\} = \{\text{Paris}\}$.

Obviously, when we use the symbol $P(x)$, we need to know whether it denotes a specific predicate or an arbitrary predicate (that is, whether P is a predicate constant or a predicate variable) just as, when we use the symbol $f(x)$ in studying numerical functions, we need to know when f is used to mean a specific function and when it is an arbitrary function of a variable x.

b. In the preceding, we have given examples of one-place predicates expressing *properties* of objects. A natural generalization of the concept of a one-place predicate is the concept of a many-place predicate, by means of which we can express *relationships* between objects.

Let us look at some examples of many-place predicates.

1. If we denote the relation "is less than," defined, for example, on the set D of real numbers, by the functional symbol "< (,)" with two blank spaces,† then we denote by the expression "< (2, 3)" the true proposition that "the number 2 is less than the number 3," and we denote by the expression "< (3, 2)" the false proposition "the number 3 is less than the number 2."

---

* As it happens, we shall not use predicate *variables* in the "restricted" predicate logic that we are studying.
† Instead of the symbol <, we can, of course, use any letter and write, for example, $R(,)$.

We denote by the expression "$< (x, y)$" the *two-place* predicate or the logical function of two numerical variables x and y which, for an arbitrary replacement of x and y by a pair of numbers in D, becomes either a true or a false proposition.*

The predicate $< (x, y)$ is defined on the set of all possible pairs of numbers belonging to D; that is, its domain of definition is the set D^2. Its range is the set $\{T, F\}$:

$$D^2 \xrightarrow{<} \{T, F\}.$$

This predicate partitions the set D^2 into two subsets on one of which it becomes a true proposition and on the other a false proposition. The subset of D^2 on which it becomes a true proposition is its truth set, which we denote by the expression

$$\{(x, y) : < (x, y)\}.$$

Suppose, instead, that the predicate $< (x, y)$ is defined on the set A^2, where $A = \{3, 4, 5, 6\}$. Since A^2 is a finite set, we can set up a table of values of the predicate $< (x, y)$ corresponding to all possible combinations of values of the variables x and y (all possible elements of A^2):

y x	3	4	5	6
3	F	T	T	T
4	F	F	T	T
5	F	F	F	T
6	F	F	F	F

Thus,

$$\{(x, y) : (x, y) \in A^2 \wedge < (x, y)\}$$
$$= \{(3, 4), (3, 5), (3, 6), (4, 5), (4, 6), (5, 6)\}.$$

In mathematics, two-place predicates (binary relations) $R(x, y)$ are usually denoted by xRy. For example, instead of writing

---

* The notation $< (x, 2)$ or, for example, $< (3, y)$, signifies, of course, a *one-place* predicate ("being less than 2" in the first case or "being greater than 3" in the second). Cf. footnote on p. 142.

$< (x, y)$, we write $x < y$; instead of writing $= (x, y)$, we write $x = y$; and instead of writing $> (x, y)$, we write $x > y$. In what follows, we shall use this customary notation for two-place predicates.

The predicate $x < y$ that we have been considering is usually referred to in mathematics as an inequality (in two unknowns) and its truth set is called the **solution set** of that inequality (on the set D if the predicate is defined on D^2 or on the set A if it is defined on A^2).

If in the predicate $x < y$ we replace one variable, let us say y, by one of its values, we obtain a logical function of the second variable, for example, $x < 3$ (an inequality in one unknown).

2. Suppose that x, y, and z are variables for points lying on a single straight line. Let us denote by $BTW(x, y, z)$ the *three-place* predicate "the point x lies between the points y and z." When we replace the variables x, y, and z by the names of specific points (lying on a single straight line), this predicate (logical function of three individual variables) becomes a true or a false proposition.*

If L is the set of all points of the straight line, then the domain of definition of the predicate $BTW(x, y, z)$ is the set L^3 (the set of all possible triples of elements of the set L) and its range is the set $\{T, F\}$:

$$L^3 \xrightarrow{BTW} \{T, F\}.$$

Let x, y, and z denote variables for the natural numbers and let $M(x, y, z)$ denote the predicate "$x + y = z$." If we replace the variables x, y, and z by any of their values, we obtain a true or a false proposition. For example, $M(2, 3, 5)$ is the true proposition $2 + 3 = 5$, and $M(1, 7, 4)$ is the false proposition $1 + 7 = 4$.

This three-place predicate is defined on the set N^3 (where N is the set of natural numbers):

$$N^3 \xrightarrow{M} \{T, F\}.$$

* Once again, we recall that, if we replace a single variable by a fixed value we get a two-place predicate; if we replace two variables by fixed values, we get a one-place predicate.

3. In general, an n-place predicate $R(x_1, x_2, \ldots, x_n)$ (meaning "$x_1, x_2, \ldots, x_n$ are in the relation R with each other") is a logical function of n individual variables defined on the set S^n of all possible n-tuples of elements of a set S:

$$S^n \xrightarrow{R} \{T, F\}.$$

An example of an n-place predicate is the equation $a_1x_1 + a_2x_2 + \ldots + a_nx_n = 0$, where $a_1, a_2, \ldots, a_n$ are definite (fixed) real numbers and $x_1, x_2, \ldots, x_n$ are variables for real numbers. The domain of definition of this predicate is D^n and its range is included in $\{T, F\}$. (To each ordered n-tuple of real numbers is assigned a true or a false proposition.)

The truth set of this predicate is the subset of D^n on which the predicate assumes the value T, that is, the set of solutions of the equation.

EXERCISES

3.55. Determine the truth set of the predicate $P(x)$, "x is a prime number," defined on the set $A = \{1, 2, 3, 4, 5, 6, 7, 8, 9\}$.

3.56. The predicate $D(x, y)$, "x divides y," is defined on the set A^2, where $A = \{1, 2, 3, 4, 5, 6,\}$. Set up a table of values of this predicate and determine $\{(x, y) : D(x, y)\}$.

3.57. The predicate $P(x, y, z)$, "$x \cdot y = z$," is defined on the set $\{0, 1\}^3$. Compile a table of values of this predicate and determine $\{(x, y, z) : P(x, y, z)\}$ (the truth set of the three-place predicate $P(x, y, z)$).

3. OPERATIONS ON PREDICATES. QUANTIFIERS

3.1. Since predicates are logical functions, that is, since they, like propositional variables, assume the values T and F, all the operations of propositional logic are applicable to them. With the aid of these operations, we can, from elementary predicates (that is, those that cannot be divided into other predicates), formulate compound predicates, just as, in propositional logic, we can form compound propositions from elementary propositions.

Let us look at the application of the operations of propositional logic to predicates by considering the case of one-place predicates. (This does not restrict the generality of our discussion since it does not depend essentially on the number of arguments.)

Suppose that a predicate $P(x)$ is defined on a set A. Then, the predicate $\bar{P}(x)$, that is, the negation of $P(x)$, is also defined on A. Specifically, $\bar{P}(x)$ becomes a true proposition for just those values x in A for which $P(x)$ becomes a false proposition; that is, the truth set of the predicate $\bar{P}(x)$ is the complement with respect to A of the truth set of the predicate $P(x)$:

$$\{x : \bar{P}(x)\} = \overline{\{x : P(x)\}}.$$

Suppose that two predicates $P(x)$ and $Q(x)$ are defined on a set A. Then it is also possible to define on A the compound predicates

$P(x) \wedge Q(x),$ \hfill (a)
$P(x) \vee Q(x),$ \hfill (b)
$P(x) \Rightarrow Q(x),$ \hfill (c)
$P(x) \Leftrightarrow Q(x),$ \hfill (d)

where the operations $\wedge$, $\vee$, $\Rightarrow$, and $\Leftrightarrow$ have the same meaning as in propositional logic.

a. The predicate $P(x) \wedge Q(x)$ becomes a true proposition for all values $x \in A$ for which both the predicates $P(x)$ and $Q(x)$ become true propositions and only for those values; that is,

$$\{x : P(x) \wedge Q(x)\} = \{x : P(x)\} \cap \{x : Q(x)\}.$$

The truth set of the predicate $P(x) \wedge Q(x)$ is the intersection of the truth sets of the predicates $P(x)$ and $Q(x)$.

b. The predicate $P(x) \vee Q(x)$ becomes a true proposition for those values of $x \in A$ for which at least one of the predicates $P(x)$ or $Q(x)$ becomes a true proposition and only for those values; that is,

$$\{x : P(x) \vee Q(x)\} = \{x : P(x)\} \cup \{x : Q(x)\}.$$

c. The predicate $P(x) \Rightarrow Q(x)$ becomes a false proposition for those values $x \in A$ for which $P(x)$ becomes a true proposition and $Q(x)$ becomes a false one and only for those values. For all remaining values of $x \in A$, the predicate $P(x) \Rightarrow Q(x)$ becomes a true proposition. The predicate $\bar{P}(x) \vee Q(x)$ assumes the same values for these values of x. (This predicate becomes a false proposition for those values of x for which $P(x)$ is a true proposition and $Q(x)$ is a false one and for only those values, and it becomes a true proposition for all other values of $x \in A$.)

Consequently,*

$P(x) \Rightarrow Q(x)$ equiv $\bar{P}(x) \vee Q(x)$

and

$\{x : P(x) \Rightarrow Q(x)\} = \{x : \bar{P}(x) \vee Q(x)\}$
$= \{x : \bar{P}(x)\} \cup \{x : Q(x)\} = \overline{\{x : P(x)\}} \cup \{x : Q(x)\}.$

d. The predicate $P(x) \Leftrightarrow Q(x)$ becomes a true proposition for those values of $x \in A$ for which $P(x)$ and $Q(x)$ both become true propositions or both become false propositions and only for such values.

Since the operations have the same meaning as in propositional logic, all their properties are maintained. For example,

$P(x) \Leftrightarrow Q(x)$ equiv $(P(x) \Rightarrow Q(x)) \wedge (Q(x) \Rightarrow P(x))$.

(Of course, this can be established anew for predicates as we have done in c.)

We also have

$P(x) \Leftrightarrow Q(x)$ equiv $(\bar{P}(x) \vee Q(x)) \wedge (\bar{Q}(x) \vee P(x))$
  equiv $(\bar{P}(x) \wedge \bar{Q}(x)) \vee (P(x) \wedge Q(x))$.
$\{x : P(x) \Leftrightarrow Q(x)\}$
$= (\overline{\{x : P(x)\}} \cup \{x : Q(x)\}) \cap (\overline{\{x : Q(x)\}} \cup \{x : P(x)\})$
$= (\overline{\{x : P(x)\}} \cap \overline{\{x : Q(x)\}}) \cup (\{x : P(x)\} \cap \{x : Q(x)\}).$

---

* Here, "equiv" has the same meaning as in the algebra of propositions; the concept of a formula in predicate logic will be made precise in the following section.

Therefore,

In propositional logic, for an arbitrary function

$\{T, F\}^n \xrightarrow{f} \{T, F\}$

(logical function of logical variables), we can construct a complete table of values (truth table). In predicate logic, if the domain of definition of the predicates is an infinite set, it is of course impossible to construct such a (finite) table. If the domain of definition is a finite set, then, although such a table can be constructed, it is extremely laborious to do so even in the simplest cases.

We shall illustrate this with an example of a truth table for the predicate $P(x) \Rightarrow Q(y)$ when the domain of definition A of the predicates P and Q consists of only two elements: $A = \{a, b\}$: that is, $x, y \in \{a, b\}$.

The composite predicate $P(x) \Rightarrow Q(y)$ is a logical function of the individual variables x and y and the predicate variables P and Q, and these in turn are logical functions of the individual variables x and y. In other words, the predicate $P(x) \Rightarrow Q(y)$ is a composite function of the variables x and y.

The values of the predicate variables P and Q are all possible functions defined on the set $\{a, b\}$ with values in the set $\{T, F\}$. All told, there are four such functions:

x	f_1	f_2	f_3	f_4
a	T	T	F	F
b	T	F	T	F

Since there are sixteen combinations of values of the predicate variables P and Q and four combinations of values of the two individual variables x and y, there are sixty-four cases for which we must indicate the value of the predicate

$P(x) \Rightarrow Q(y)$.

Let us construct a table with four columns and sixteen rows:*

* Columns 1 and 4 of this table constitute a table of values of the predicate $P(x) \Rightarrow Q(x)$ defined on the same set A.

$P(x)$	$Q(y)$	$P(a) \Rightarrow Q(a)$	$P(a) \Rightarrow Q(b)$	$P(b) \Rightarrow Q(a)$	$P(b) \Rightarrow Q(b)$
f_1	f_1	T	T	T	T
f_1	f_2	T	F	T	F
f_2	f_1	T	T	T	T
f_2	f_2	T	F	T	T
f_1	f_3	F	T	F	T
f_3	f_1	T	T	T	T
f_3	f_3	T	T	F	T
f_1	f_4	F	F	F	F
f_4	f_1	T	T	T	T
f_4	f_4	T	T	T	T
f_2	f_3	F	T	T	T
f_3	f_2	T	T	T	F
f_2	f_4	F	F	T	T
f_4	f_2	T	T	T	T
f_3	f_4	T	T	F	F
f_4	f_3	T	T	T	T

When the domain of definition A of the predicates $P(x)$ and $Q(y)$ consists of a single element $A = \{a\}$, these predicates no longer constitute a function of an individual variable since, in this case, the "variable" has only the single value a. In this case, $P(a)$ and $Q(a)$, or simply P and Q, are elementary propositions (if P and Q have been denoting predicate constants) or propositional variables (if they have been denoting predicate variables). Propositional variables can be treated as "zero-place" predicates; that is, just as we have a three-place (or two-place or one-place) predicate when $n = 3$ (or 2 or 1) that assigns to each triple (or pair or single element) of elements belonging to A a value in $\{T, F\}$, when $n = 0$ we shall assume that we have a zero-place predicate that is a variable assuming values in $\{T, F\}$.

3.2. The operations of propositional logic transform predicates into predicates. Let us now look at operations that transform predicates into propositions.

One such operation is the replacement of individual variables by their values. Consider a predicate $P(x_1, x_2, \ldots, x_n)$, where P is a predicate constant and $x_1, x_2, \ldots, x_n$ are individual variables. If we replace each of the individual variables x_i by one of its values a_i, we obtain a (true or false) proposition $P(a_1, a_2, \ldots, a_n)$, concerning a particular combination of objects $(a_1, a_2, \ldots, a_n)$.

For example, let $E(x)$ denote the proposition

x is an even number.

If we replace the variable x by the value 6, we obtain the true proposition $E(6)$, that is, the proposition

6 is an even number.

If instead we replace x by the number 5, we obtain the false proposition $E(5)$, that is, the proposition

5 is an even number.

Both these propositions relate to individual objects, namely, the numbers 6 and 5.

Consider the two-position predicate $x \colon y$,

x is an integral multiple of y.

If we replace the variables x and y by the pair* of numbers (6, 2), we obtain the true proposition $6 \colon 2$, that is, the proposition

6 is an integral multiple of 2.

If we substitute the pair (5, 2), we obtain the false proposition $5 \colon 2$, that is, the proposition

5 is an integral multiple of 2.

Each of these propositions relates to an ordered pair of numbers: (6, 2) in the first case and (5, 2) in the second.

However, it is possible to construct from predicates not only propositions relating to a definite object or ordered set (pair,

---

* In the Introduction, we agreed to use the word "pair" everywhere to mean an ordered pair.

triple, etc.) of objects but also propositions expressing a property or relationship of objects of an entire set (propositions regarding generality) and propositions on the existence of objects belonging to a given set possessing a certain property or having a certain relationship with other objects (propositions regarding existence). To construct such propositions in predicate logic, one introduces quantifiers.

Suppose that we have a certain predicate $P(x)$. It is possible that the property P is possessed by all elements in the domain of definition of this predicate or at least by some of these elements. In the first case, the proposition

For all x (in the given set), $P(x)$ holds

is true; in the second case, the proposition

There exists an x for which $P(x)$ holds

is true.

We denote the expression "for all x" (or "for every x" or "for arbitrary x") by the expression $(\forall x)$, and we call it a **universal quantifier**.

We denote the expression "there exists an x such that" by the expression $(\exists x)$ and call it an **existential quantifier**.

Writing one of these quantifiers in front of a predicate formula expresses the operation of binding with a quantifier: The variable that is "bound" by this quantifier (that is, the variable appearing both in the predicate and in the quantifier) is called a **bound variable**.

For example, if $P(x)$ is the predicate

x is a prime number,

then $(\forall x) P(x)$ is the false proposition

Every number x is a prime,

but $(\exists x) P(x)$ is the true proposition

There exists a number x such that x is a prime.

The universal quantifier can be regarded as a generalization of conjunction and the existential quantifier can be regarded as a generalization of disjunction. If the domain of definition A of a predicate $P(x)$ is finite, for example, if $A = \{a_1, a_2, \ldots, a_n\}$, then the proposition $(\forall x)\, P(x)$ is equivalent to the conjunction

$$P(a_1) \wedge P(a_2) \wedge \cdots \wedge P(a_n),$$

and the proposition $(\exists x)\, P(x)$ is equivalent to the disjunction

$$P(a_1) \vee P(a_2) \vee \cdots \vee P(a_n).$$

On the other hand, if the predicate $P(x)$ is defined on an infinite set, the quantifiers play the role of "infinite" conjunctions and disjunctions, in much the same way as infinite series and integrals are generalizations of ordinary finite sums in mathematical analysis.

In the language of predicate logic, we can analyze arguments of the types 1 and 2 introduced at the beginning of Section 2.1 as illustrations of the inadequacy of propositional logic for such an analysis.

Let us look at 1:

every integer is a rational number,
1 is an integer;
―――――――――――――――――――――
consequently, 1 is a rational number.

We introduce two one-place predicates defined, for example, on the set D of real numbers:

$C(x)$: x is an integer,

and

$R(x)$: x is a rational number.

By using these predicates and a universal quantifier, we can formulate the first premise as follows: "For every x, if x is an integer, then x is a rational number." Consequently, we can write this as follows in the notation adopted:

$(\forall x)[C(x) \Rightarrow R(x)]$.

The second premise can be written as $C(1)$ and the conclusion as $R(1)$.

Let us now establish the validity of the reasoning of Type 1; that is, let us show that, if the premises are valid, the conclusion cannot be false. Since the proposition $(\forall x)[C(x) \Rightarrow R(x)]$ is a true proposition, the predicate $C(x) \Rightarrow R(x)$ becomes a true proposition when we replace the variable x by any one of its values:

x	$C(x)$	$R(x)$	$C(x) \Rightarrow R(x)$
$\sqrt{2}$	F	F	T
0.5	F	T	T
—3	T	T	T

Let us replace x by the value 1. We obtain the true proposition $C(1) \Rightarrow R(1)$. In accordance with the detachment rule (*modus ponens*) it follows from the truth of the propositions $C(1) \Rightarrow R(1)$ and $C(1)$ that $R(1)$ is also a true proposition.

If we interpret C and R as predicate variables and 1 as the name of any element in the domain of definition of the predicates $C(x)$ and $R(x)$, we have established the validity not only of the single specific argument (1) but also any other argument having the same structure:

$(\forall x)[C(x) \Rightarrow R(x)]$,
$\underline{C(1);}$
consequently, $R(1)$.

In particular, we have also established the validity of the argument in Type 2:

every rhombus is a parallelogram,
$\underline{ABCD \text{ is a rhombus;}}$
consequently, $ABCD$ is a parallelogram.

In this case, the predicate variables $C(x)$ and $R(x)$ assume respectively the following values:*
x is a rhombus,
x is a parallelogram.

The operation of binding with a quantifier or, as it is called, **quantification**, converts a one-place predicate into a proposition. To convert a many-place predicate into a proposition by means of quantification, we need to put as many quantifiers in front of it as there are distinct variables, thus binding each individual variable by means of a quantifier.

Suppose that we have a two-place predicate $R(x, y)$. With the aid of quantifiers, we can construct from it the following propositions:

1. $(\forall x)(\forall y)R(x, y)$ "for every x and every y, the proposition $R(x, y)$ holds";
2. $(\forall y)(\forall x)R(x, y)$ "for every y and every x, the proposition $R(x, y)$ holds";
3. $(\forall x)(\exists y)R(x, y)$ "for every x, there exists a y such that $R(x, y)$ holds";
4. $(\exists y)(\forall x)R(x, y)$ "there exists a y such that, for every x, the proposition $R(x, y)$ holds";

---

* As was noted previously (cf. footnote p. 140), the language of this "restricted" predicate logic that we are considering (and of the "restricted" predicate *calculus* discussed superficially in the following pages) contains no "variables" in the literal sense of the word, that is, no symbols that can be replaced, in accordance with specific rules, by the names of specific predicates from some class and that can be bound by quantifiers. This is in contrast with the "extended" predicate calculus considered, for example, in [3], Chapter 4, in which there are expressions of the form $(\forall F)F(x)$ and $(\exists F)F(x)$, where $F(x)$ is a predicate variable; of course, the combinations of symbols of the form $(\forall A)A(x)$ or $(\exists A)A(x)$, where $A(x)$ denotes some *specific* predicate, are meaningless just as, for example, expressions of the form $(\forall 1)$ or $(\exists 0)$ are meaningless in predicate logic.

The abbreviations and simplifications in our phraseology that we have used, like those just used in the text, actually indicate that, in the given context, the symbols $C(x)$ and $R(x)$ will be used as notations for the *specific* predicates "x is a rhombus" and "x is a parallelogram," respectively. When we wish to make some assertion dealing with an arbitrary predicate, we can use for this purpose *metalinguistic* symbols, which are to be understood as symbols for *arbitrary* predicates.

5. $(\exists x)(\forall y)R(x, y)$ "there exists an x such that, for every y, the proposition $R(x, y)$ holds";
6. $(\forall y)(\exists x)R(x, y)$ "for every y, there exists an x such that $R(x, y)$ holds";
7. $(\exists x)(\exists y)R(x, y)$ "there exists an x and there exists a y such that $R(x, y)$ holds";
8. $(\exists y)(\exists x)R(x, y)$ "there exists a y and there exists an x such that $R(x, y)$ holds."

If a two-place predicate is bound by a single quantifier, for example, $(\forall x)R(x, y)$, then the resulting formula expresses not a proposition but a logical function of the second variable, which is not bound by the quantifier (and we say that it is a **free** variable). This formula is a one-place predicate.

Suppose, for example, that x and y are variables for real numbers (in which case we usually say "x and y are real numbers") and suppose that $<$ is the symbol for the two-place predicate (relation) that we know as "less than." Then,

$(\forall x)(\forall y)[x < y]$ and $(\forall y)(\forall x)[x < y]$	are false propositions;
$(\forall x)(\exists y)[x < y]$	is a true proposition, but
$(\exists y)(\forall x)[x < y]$	is a false proposition;
$(\exists x)(\forall y)[x < y]$	is a false proposition, but
$(\forall y)(\exists x)[x < y]$	is a true proposition;
$(\exists x)(\exists y)[x < y]$ and $(\exists y)(\exists x)[x < y]$	are true propositions;
$(\forall x)[x < y]$	is a logical function of y;
$(\exists y)[x < y]$	is a logical function of x.

3.3. Let us give some examples of the expression of mathematical propositions in the language of predicate logic.

a. The proposition "every natural number is greater than zero" can be formulated in the form of an implication with a universal quantifier:

Every number, if it is a natural number, is greater than zero.

We can write it as follows:

$(\forall x)[(x \in N) \Rightarrow (x > 0)]$.

This expression can be somewhat simplified (in form, of course, not in substance) if, instead of the universal quantifier "for all x," we use a so-called **restricted** universal quantifier "for all x belonging to N," which we write in symbols as follows:

$(\forall x)_{\in N}$.

With the restricted quantifier, the proposition shown above may be written

$(\forall x)_{\in N}[x > 0]$.

b. The proposition "lim $a_n = l$," meaning "the limit of the sequence $\{a_n\}$ is the number l" means by definition "for every positive number ε, there exists a natural number n_ε such that, for every natural number n, if $n > n_\varepsilon$, then $|a_n - l| < \varepsilon$."*

Let us write this definition in the language of predicate logic using restricted quantifiers:

$[\lim a_n = l] \text{ equiv}_{Df} (\forall \varepsilon)_{>0} (\exists n_\varepsilon)_{\in N} (\forall n)_{\in N} [(n > n_\varepsilon) \Rightarrow (|a_n - l| < \varepsilon)]$.

c. Let us write in the language of predicate logic the definition of continuity of a function at a point.

Suppose that the domain of definition M_f of a function f is an interval containing the point x_0. In this case, the definition of continuity of the function f at the point x_0 can be formulated as follows:

The function f is continuous at the point x_0 if and only if, for every positive number ε, there exists a positive number δ such that, for every x in the domain of definition M_f of the function f,

if $|x - x_0| < \delta$, then $|f(x) - f(x_0)| < \varepsilon$.

$\begin{bmatrix} f \text{ is continuous} \\ \text{at the point } x_0 \end{bmatrix}$ equiv$_{Df}$ $(\forall \varepsilon)_{>0} (\exists \delta)_{>0} (\forall x)_{\varepsilon M_f} [(|x - x_0| < \delta) \Rightarrow (|f(x) - f(x_0)| < \varepsilon)]$.

* The formulation of this definition in words is such that we can write it in the symbols of predicate logic without making any transformations.

d. Let us look at the geometric axiom:

For any two points, there exists a straight line passing through them.

We introduce the following predicates:

$T(x)$: x is a point,
$P(x)$: x is a straight line,
$I(x, y)$: x passes through y.

With these predicates, our axiom can be written in the following form:

$$(\forall x)(\forall y)[T(x) \wedge T(y) \Rightarrow (\exists z)(P(z) \wedge I(z, x) \wedge I(z, y))].$$

EXERCISES

3.58. Consider the following predicates:

$N(x)$: x is a natural number,
$C(x)$: x is an integer,
$P(x)$: x is a prime number,
$\Pi(x)$: x is a positive number,
$E(x)$: x is an even number,
$D(x, y)$: x divides y.

Write in English the propositions written below in the symbols of predicate logic, and indicate which are true and which are false:

a. $(\forall x)[N(x) \Rightarrow C(x)]$;
b. $(\exists x)[N(x) \wedge C(x)]$;
c. $(\forall x)[C(x) \Rightarrow N(x)]$;
d. $(\forall x)[C(x) \wedge \Pi(x) \Leftrightarrow N(x)]$;
e. $(\forall x)[C(x) \Rightarrow E(x) \vee \bar{E}(x)]$;
f. $(\forall x)(\exists y)[C(x) \wedge C(y) \Rightarrow D(x, y)]$;
g. $(\exists y)(\forall x)[C(x) \wedge C(y) \Rightarrow D(x, y)]$;
h. $(\exists x)(\forall y)[C(x) \wedge C(y) \Rightarrow D(x, y)]$;
i. $(\forall x)(\forall y)[E(x) \wedge \bar{E}(y) \Rightarrow \bar{D}(x, y)]$;
j. $(\exists x)[P(x) \wedge E(x)]$;
k. $(\forall x)[P(x) \Rightarrow \bar{E}(x)]$.

3.59. Consider the following predicates:

$C(x)$: x is a composite number;
$R(x, y)$: x is less than y;
$S(x, y, z)$: $x + y = z$;
$P(x, y, z)$: $x \cdot y = z$.

Using the notation given above, write the following propositions using the symbolism of predicate logic:

     a. For every number x and every number y, there exists a number z such that $x + y = z$.

     b. For every x, y, and z, if $x + y = z$, then $y + x = z$.

     c. For every number x and every number y, there exists a number z such that $x \cdot y = z$.

     d. For every x, y, and z, if $x \cdot y = z$, then $y \cdot x = z$.

     e. For every number x and every number z, there exists a number y such that $x + y = z$.

     f. For every number x distinct from zero and every number z, there exists a number y such that $x \cdot y = z$.

     g. For every x and y, the number x is less than y if and only if there exists a positive number k such that $x + k = y$.

     h. For every x, x is composite if and only if there exist numbers u and v less than x such that $u \cdot v = x$.

4. FORMULAS OF PREDICATE LOGIC. EQUIVALENT FORMULAS. UNIVERSALLY VALID FORMULAS

4.1. The concept of a formula in predicate logic can be defined as follows:

     1. All formulas of propositional logic are formulas of predicate logic.

     2. Every predicate symbol is a formula when the blank spaces in it are occupied by individual variables or constants.

     3. If $\varphi(X_1, X_2, \ldots, X_n)$ is a formula of propositional logic, then the result of replacing the propositional variables $X_1, X_2, \ldots, X_n$ by arbitrary formulas of predicate logic $\varphi_1, \varphi_2, \ldots, \varphi_n$ will also be a formula of predicate logic under the following condition: if any individual variable is free in one of the formulas φ_i, it is

not bound in any of the others. In the resulting formula $\varphi(\varphi_1, \varphi_2, \ldots, \varphi_n)$, the individual variables are free or bound according as they are free or bound in the formulas $\varphi_1, \varphi_2, \ldots, \varphi_n$.

4. If $\varphi(\cdots x_i \cdots)$ is a formula,* then $(\forall x_i)\ \varphi(\cdots x_i \cdots)$ and $(\exists x_i)\ \varphi(\cdots x_i \cdots)$ are also formulas. In these formulas, the variable x_i is bound and the other variables are free or bound according as they are free or bound in the formula $\varphi(\cdots x_i \cdots)$.

5. There are no other formulas of predicate logic.

Now, we can determine for an arbitrary given finite sequence of symbols of predicate logic whether this sequence constitutes a formula or not.

For example, let us show that the sequence of symbols

$$(\exists x)(\forall y)P(x, y) \vee (\forall z)S(z) \wedge A$$

is a formula.

1. A is a formula (in accordance with Clause 1 of the definition).
2. $P(x, y)$ and $S(z)$ are formulas (in accordance with Clause 2 of the definition).
3. $(\forall y)\ P(x, y)$ and $(\forall z)\ S(z)$ are formulas (as a consequence of Step 2 in accordance with Clause 4 of the definition).
4. $(\exists x)\ (\forall y)\ P(x, y)$ is a formula (as a consequence of Step 3 in accordance with Clause 4 of the definition).
5. $(\exists x)\ (\forall y)\ P(x, y) \vee (\forall z)\ S(z) \wedge A$ is a formula (as a consequence of steps 1, 3, and 4 in accordance with Clause 3 of the definition, replacing A_1 in the formula $A_1 \vee A_2 \wedge A$ of propositional logic by the formula $(\exists x)(\forall y)\ P(x, y)$ and replacing A_2 by the formula $(\forall z)\ S(z)$).

The elementary components of this (as of any) formula of predicate logic are elementary (nondecomposable) predicates, including zero-place predicates (propositions). In the present

---

* We denote by $\varphi(\ldots x_i \ldots)$ an arbitrary formula of predicate logic in which the individual variable x_i appears free (that is, is unbound by quantifiers). (Here, the dots stand for other individual variables that may appear in φ.) The requirement that x_i appear free in φ is not obligatory: we may assume that, for a predicate formula φ that is independent of x, $(\forall x)\ \varphi$ and $(\exists x)\ \varphi$ simply coincide with φ.

formula, these are the two-place predicate $P(x, y)$, the one-place predicate $S(z)$, and the "zero-place predicate" A.

Every formula of predicate logic expresses a logical function of the predicate variables and free individual variables contained in it. Since there are no free individual variables in the formula given above (all the individual variables x, y, and z are bound by the quantifiers), this formula expresses a logical function of the predicate variables P, S, and A.

4.2. If two formulas φ_1 and φ_2 express the same logical function of the predicate variables and free individual variables, that is, if they both assume the value T or both the value F for an arbitrary combination of values of these variables,* they are said to be **equivalent**. This phenomenon is expressed, as in propositional logic, by writing φ_1 equiv φ_2 (or by using one of the symbols $=$, $\equiv$, or $\sim$).

This definition of the equivalence of formulas of predicate logic is a generalization of the definition of the equivalence of formulas of propositional logic. If φ_1 and φ_2 do not contain predicate variables but only variables for the "zero-place predicates," that is, only propositional variables, then φ_1 and φ_2 are equivalent formulas of propositional logic. Thus, equivalent formulas of propositional logic are equivalent formulas of predicate logic.

Let us look at some equivalences of formulas of predicate logic that generalize certain equivalences of formulas of propositional logic to the case of "infinite" conjunctions and disjunctions.

a. If the domain of definition of the predicate $P(x)$ is a finite set $A = \{a_1, a_2, \ldots, a_n\}$, it follows from the meanings of the universal and existential quantifiers that

$$(\forall x)P(x) \text{ equiv } \bigwedge_{i=1}^{n} P(a_i)$$

and

$$(\exists x)P(x) \text{ equiv } \bigvee_{i=1}^{n} P(a_i).$$

---

* The values of the predicate variables are specific predicates defined on some set, and the values of the free individual variables are elements of that set.

Therefore, we also have*

$$\overline{(\forall x)P(x)} \text{ equiv } \overline{\bigwedge_{i=1}^{n} P(a_i)},$$

$$\overline{(\exists x)P(x)} \text{ equiv } \overline{\bigvee_{i=1}^{n} P(a_i)}.$$

By using de Morgan's laws (see Chapter 1, p. 60, Laws 16 and 17), generalized to the case of n propositions, we find

$$\overline{\bigwedge_{i=1}^{n} P(a_i)} \text{ equiv } \bigvee_{i=1}^{n} \bar{P}(a_i)$$

and

$$\overline{\bigvee_{i=1}^{n} P(a_i)} \text{ equiv } \bigwedge_{i=1}^{n} \bar{P}(a_i),$$

and by the transitivity of the relation of equivalence (see Chapter 1, Section 2.3), we obtain

$$\overline{(\forall x)P(x)} \text{ equiv } \overline{(\exists x)\bar{P}(x)},$$
$$\overline{(\exists x)P(x)} \text{ equiv } \overline{(\forall x)\bar{P}(x)}.$$

These equivalences remain valid for the case in which the domain of definition of $P(x)$ is an infinite set, as a generalization of de Morgan's laws to the case of "infinite" conjunctions and disjunctions. They are in accord with the ordinary meaning of the quantifiers. The proposition

It is not true that every x (in a given set) possesses property P

means in ordinary language the same thing as the proposition

There exists an x (in the given set) that does not possess property P (or that possesses the property $\bar{P}$).

This is what is asserted in Equivalence 1.

* We write $\overline{(\forall x)}P(x)$ instead of $\overline{(\forall x)P(x)}$ and $\overline{(\exists x)}P(x)$ instead of $\overline{(\exists x)P(x)}$

SEC. 4. FORMULAS OF PREDICATE LOGIC 159

The proposition

It is not true that there exists an x (in the given set) that possesses property P

means the same thing as the proposition

There is no x (in the given set) possessing property P

or

Every x possesses the property $\bar{P}$.

This is what is asserted in Equivalence 2.*

From equivalences 1 and 2, we obtain the following rule of transformation of a negation of a proposition with a quantifier: The universal quantifier is replaced by the existential quantifier and vice versa and the negation symbol is transferred to the expression following the quantifier.

This rule is also applicable to formulas with several quantifiers. For example,

$$\overline{(\forall x)(\exists y)P(x, y)} \text{ equiv } (\exists x)\overline{(\exists y)P(x, y)}$$
$$\text{equiv } (\exists x)(\forall y)\bar{P}(x, y).$$

If $x, y, \ldots$ are variables for natural numbers and $P(x, y)$ is the predicate "x is greater than y," then $\overline{(\forall x)(\exists y)P(x, y)}$ denotes the proposition

It is not true that, for an arbitrary natural number x, there exists a natural number y such that x is greater than y,

and $(\exists x)(\forall y)\bar{P}(x, y)$ denotes the equivalent proposition

There exists a natural number x such that, for every natural number y, x is not greater than y.

---

* In this and the following section, we use what is known as the law of the excluded middle ("every object either possesses property P or fails to possess property P"), the use of which in this book is nowhere restricted by us and is not explicitly mentioned. We have already assumed this law in propositional logic in the guise of Formula 14, p. 60. For more detail, see, for example, [5], Section 13.

As an example, let us apply the rule of transformation of a negation to translate into the language of predicate logic the proposition*

The function f is not continuous at the point x_0.

This proposition is the negation of the proposition

The function f is continuous at the point x_0,

which, by definition, means

$$(\forall \varepsilon)(\exists \delta)(\forall x)[(|x - x_0| < \delta) \Rightarrow (|f(x) - f(x_0)| < \varepsilon)].$$
$$_{>0 \quad >0 \quad \varepsilon M_f}$$

The negation of this formula is

$$\overline{(\forall \varepsilon)(\exists \delta)(\forall x)[(|x - x_0| < \delta) \Rightarrow (|f(x) - f(x_0)| < \varepsilon)]}.$$
$$_{>0 \quad >0 \quad \varepsilon M_f}$$

We can transform it into

$$(\exists \varepsilon)(\forall \delta)(\exists x)[\overline{(|x - x_0| < \delta \Rightarrow (|f(x) - f(x_0)| < \varepsilon)}],$$
$$_{>0 \quad >0 \quad \varepsilon M_f}$$

or $(\exists \varepsilon)(\forall \delta)(\exists x)[(|x - x_0| < \delta) \wedge (|f(x) - f(x_0)| \geq \varepsilon],$
$_{>0 \quad >0 \quad \varepsilon M_f}$

that is, the proposition

The function f is not continuous at the point x_0

means the same thing as the proposition

There exists a positive number ε such that, for every positive number δ, there exists an x in the domain of definition of the function f such that $|x - x_0| < \delta$ and at the same time, $|f(x) - f(x_0)| \geq \varepsilon$.

b. We have the equivalence

$(\forall x)P(x) \wedge (\forall x)Q(x)$ equiv $(\forall x)[P(x) \wedge Q(x)],$

which is a generalization of the properties of commutativity and associativity of conjunction to the case of "infinite conjunction."

* It is assumed, just as above, that f is defined on an interval containing x_0.

To prove this equivalence, note that, if the predicates $P(x)$ and $Q(x)$ are defined on a finite set $\{a_1, a_2, \ldots, a_n\}$, then

$$(\forall x)P(x) \wedge (\forall x)Q(x) \text{ equiv } \bigwedge_{i=1}^{n} P(a_i) \wedge \bigwedge_{i=1}^{n} Q(a_i)$$

$$\text{equiv } \bigwedge_{i=1}^{n} (P(a_i) \wedge Q(a_i))$$

$$\text{equiv } (\forall x)[P(x) \wedge Q(x)].$$

The formulas $(\forall x)P(x) \vee (\forall x)Q(x)$ and $(\forall x)[P(x) \vee Q(x)]$ are not equivalent. For example, if $P(x)$ is the predicate "x is an even number" and $Q(x)$ is the predicate "x is an odd number," then the first of these expresses a false proposition, whereas the second expresses a true proposition.

Thus, the operation of binding with the universal quantifier is distributive with respect to conjunction but not disjunction.

c. We have the equivalence

$$(\exists x)P(x) \vee (\exists x)Q(x) \text{ equiv } (\exists x)[P(x) \vee Q(x)].$$

This equivalence is a generalization of the properties of commutativity and associativity for "infinite disjunction."

The formulas $(\exists x)P(x) \wedge (\exists x)Q(x)$ and $(\exists x)[P(x) \wedge Q(x)]$ are not equivalent. For example, with the same values of the predicates $P(x)$ and $Q(x)$ as in Example b, the first of these formulas expresses a true proposition and the second a false one.

Thus, the operation of binding with the existential quantifier is distributive with respect to disjunction but not with respect to conjunction.

4.3. If two equivalent formulas of predicate logic are connected by the symbol ⇔, then the resulting formula assumes the value T for an arbitrary combination of values of the predicate variables and free individual variables in an arbitrary domain.

To see this, note that, if φ_1 equiv φ_2, then φ_1 and φ_2 both assume the value T or both the value F for an arbitrary combination of values of their predicate variables and free individual variables.

Consequently, the formula

$\varphi_1 \Leftrightarrow \varphi_2$

always assumes the value T.

For example, the formula $(\overline{\forall x})P(x) \Leftrightarrow (\exists x)\bar{P}(x)$ assumes the value T for an arbitrary value of the predicate variable P and an arbitrary domain of definition of the predicate $P(x)$.

The formula $P(x, y) \vee Q(x, y) \Leftrightarrow Q(x, y) \vee P(x, y)$ assumes the value T for arbitrary values of the predicate variables P and Q (in any domain of definition of these predicates), and for arbitrary values of the free individual variables in that domain.

A formula in predicate logic that assumes the value T for arbitrary values of its predicate variables in an arbitrary domain of definition of these predicates and for arbitrary values of the free individual variables in the domain of definition is called a **universally valid** formula.

The proposition "φ is a universally valid formula" is denoted by $\vDash \varphi$. (The symbol $\vDash$ is a metalinguistic symbol used to shorten the writing of a proposition about a formula in the language of predicate logic, namely, the proposition that this formula possesses the property of universal validity.) The concept of universal validity of a formula in predicate logic is a generalization of the concept of a tautology in the algebra of propositions: All tautologies in propositional logic are universally valid formulas in predicate logic. In particular, this holds for formulas 1–30 of Section 2 of Chapter 1. These formulas serve as a source of new universally valid formulas, which can be obtained by replacing the letters in formulas 1–30 by arbitrary formulas of predicate logic.

For example, if we replace X by $P(x_1, x_2, \ldots, x_n)$ in Formula 14, that is, in the formula $X \vee \bar{X}$ (the law of the excluded middle), we obtain the universally valid formula

$P(x_1, x_2, \ldots, x_n) \vee \bar{P}(x_1, x_2, \ldots, x_n)$.

To see this, let P^0 denote an arbitrary n-place predicate defined on a set S and let $(a_1, a_2, \ldots, a_n)$ denote an arbitrary n-tuple of values of the individual variables in S. If we replace the variables by these values, we obtain the proposition

$$P^0(a_1, a_2, \ldots, a_n) \lor \overline{P^0}(a_1, a_2, \ldots, a_n),$$

which, in accordance with Formula 14, has the value T.

Since P^0 is an arbitrary value of the predicate variable P and $(a_1, a_2, \ldots, a_n)$ is an arbitrary set of values of the individual variables $x_1, x_2, \ldots, x_n$, we have

$$\vDash P(x_1, x_2, \ldots, x_n) \lor \bar{P}(x_1, x_2, \ldots, x_n).$$

The universally valid formula previously given

$$P(x, y) \lor Q(x, y) \Leftrightarrow Q(x, y) \lor P(x, y)$$

is obtained from Formula 2, that is, from the formula $X \lor Y \Leftrightarrow Y \lor X$ by replacing X by the formula $P(x, y)$ and replacing Y by the formula $Q(x, y)$.

However, not all universally valid formulas of predicate logic can be obtained by this procedure from formulas 1–30. For example, we cannot obtain the formula

$$(\overline{\forall x})P(x) \Leftrightarrow (\exists x)\bar{P}(x)$$

by this procedure from formulas 1–30.

Universally valid formulas express the laws of logic in the language of the predicate logic.

As a supplement to the universally valid formulas 1–30, we give below a list of certain other frequently used universally valid formulas of predicate logic:

31. $(\overline{\forall x})P(x) \Leftrightarrow (\exists x)\bar{P}(x)$;
32. $(\overline{\exists x})P(x) \Leftrightarrow (\forall x)\bar{P}(x)$;
33. $(\forall x)P(x) \Leftrightarrow (\overline{\exists x})\bar{P}(x)$;
34. $(\exists x)P(x) \Leftrightarrow (\overline{\forall x})\bar{P}(x)$;
35. $(\forall x)[P(x) \land Q(x)] \Leftrightarrow (\forall x)P(x) \land (\forall x)Q(x)$;
36. $(\exists x)[P(x) \lor Q(x)] \Leftrightarrow (\exists x)P(x) \lor (\exists x)Q(x)$;
37. $(\forall x)(\forall y)P(x, y) \Leftrightarrow (\forall y)(\forall x)P(x, y)$;
38. $(\exists x)(\exists y)P(x, y) \Leftrightarrow (\exists y)(\exists x)P(x, y)$;

39. $(\exists x)(\forall y)P(x, y) \Rightarrow (\forall y)(\exists x)P(x, y)$;
40. $(\forall x)P(x) \lor (\forall x)Q(x) \Rightarrow (\forall x)[P(x) \lor Q(x)]$;
41. $(\exists x)[P(x) \land Q(x)] \Rightarrow (\exists x)P(x) \land (\exists x)Q(x)$;
42. $(\forall x)[P(x) \Rightarrow Q(x)] \Rightarrow ((\forall x)P(x) \Rightarrow (\forall x)Q(x))$;
43. $(\forall x)P(x) \Rightarrow (\exists x)P(x)$;
44. $(\forall x)P(x) \Rightarrow P(y)$;
45. $P(y) \Rightarrow (\exists x)P(x)$.

Formulas 31, 32, 35, and 36 were obtained in Section 4.2 of this chapter in a somewhat different form (in the form of an equivalence of two formulas, φ_1 equiv φ_2, instead of $\varphi_1 \Leftrightarrow \varphi_2$), as generalizations of de Morgan's laws and of the commutativity and associativity of conjunction and disjunction.

Formulas 33 and 34 are easily obtained from formulas 31 and 32, respectively. Specifically, if we write Formula 31 in the form

$(\overline{\forall x})P(x)$ equiv $(\exists x)\bar{P}(x)$.

and take the negations of both formulas, we obtain

$(\overline{\overline{\forall x}})P(x)$ equiv $(\overline{\exists x})\bar{P}(x)$.

By applying the law of double negation, we obtain

$(\forall x)P(x)$ equiv $(\overline{\exists x})\bar{P}(x)$,

or

$\vdash (\forall x)P(x) \Leftrightarrow (\overline{\exists x})\bar{P}(x)$.

In an analogous manner, we obtain Formula 34 from Formula 32.

Formulas 33 and 34, like formulas 31 and 32, reflect the ordinary meaning of the quantifiers. When we say, for example, that "every x (in a given set) possesses property P," we mean by this the same thing as when we say "there does not exist an x (in the given set) that does not possess property P."*

Formulas 37 and 38 can also be regarded as generalizations of the properties of commutativity and associativity to the case

* Cf. footnote on p. 160.

of "infinite" disjunction (37) and conjunction (38). If the domain of definition of the predicate $P(x, y)$ is a finite set, these formulas can be reduced to formulas of propositional logic.

Formulas 39–45 are implications. One can easily show that the converse implications* are not universally valid. To do this, we need only find one domain (infinite or finite) and one combination of values in that domain of the predicate variables and free individual variables such that the formula assumes the value F.

Suppose, for example, that x and y are variables for real numbers and that $P(x, y)$ is the predicate "x is less than y." Then, the formula $(\forall y)(\exists x)P(x, y)$, that is, the formula "for every y, there exists an x such that $x < y$," is a true proposition; that is, it assumes the value T. However, the formula $(\exists x)(\forall y)P(x, y)$, that is, the formula "there exists an x such that $x < y$ for every y," assumes the value F (since there is no least number in the set of real numbers).

Therefore, the converse implication to Formula 39:

$$(\forall y)(\exists x)P(x, y) \Rightarrow (\exists x)(\forall y)P(x, y)$$

has the value F and thus is not universally valid. From this it follows that the formulas $(\exists x)(\forall y)P(x, y)$ and $(\forall y)(\exists x)P(x, y)$ are not equivalent.

Thus, in the transformation of formulas, we may not interchange two quantifiers of opposite types though, by virtue of formulas 37 and 38, we can interchange quantifiers of the same type.

4.4. The fundamentals of predicate logic have been described above (Sections 2–4) in terms of the meaning of the symbols. However, like propositional logic, predicate logic can be constructed in a purely formal way as an axiomatic system, i.e. as a predicate calculus.

This construction is carried out along the same lines as the construction of the propositional calculus described in Chapter 2; that is, an alphabet is given and the notion of **formula** of the

---

* For an implication $\varphi_1 \Rightarrow \varphi_2$, the converse implication is the implication $\varphi_2 \Rightarrow \varphi_1$.

predicate calculus is defined. Then, from the set of formulas, we single out the subset of derivable formulas by indicating the initial derivable formulas (axioms) and the rules of inference with the aid of which new derivable formulas can be obtained from the axioms.

In particular, we can take as the system of axioms of the predicate calculus the system of axioms presented in Chapter 2, Section 2, for the propositional calculus, supplemented with the two axioms*

44. $(\forall x)P(x) \Rightarrow P(y)$
45. $P(y) \Rightarrow (\exists x)P(x)$

and the two rules of inference

$$\frac{Q \Rightarrow P(x)}{Q \Rightarrow (\forall x)P(x)}$$

and

$$\frac{P(x) \Rightarrow Q}{(\exists x)P(x) \Rightarrow Q}$$

In both rules of inference, Q is a formula that does not contain x as a free variable.

This system of axioms is consistent, independent, and complete in the broad sense, though not complete in the narrow sense.† Every derivable formula of the predicate calculus is a universally valid formula of the contensive predicate logic and vice versa.†

The decision problem for the predicate calculus (the question as to the existence of a method enabling us to decide whether an arbitrary specific formula of the predicate calculus is derivable) is answered in the negative.‡ (It is answered affirmatively only for

---

* In axioms 44 and 45, y is a variable. In more extensive mathematical theories constructed on the basis of the predicate calculus (for example, in arithmetic, see pp. 181 and 188), we can also let y be any individual constant.
† See [8], Chapter 4.
‡ Since the derivable formulas coincide with the universally valid formulas, the decision problem is concerned with the existence of a method enabling us to decide whether any given formula of the predicate calculus is universally valid. A negative solution to this problem was given by Alonzo Church in 1936. For a proof of Church's Theorem, cf. Kleene [5], Chapter XIV, Section 76, or Mendelson [6], Chapter 3, Section 6.

special classes of formulas, in particular, formulas containing only one-place predicates.)*

EXERCISES

3.60. Determine whether the following sequences of symbols are formulas of predicate logic:

a. $(\forall x)P(x, y) \Rightarrow (\forall z)Q(y, z) \land R(y, u)$;
b. $(\forall x)A(x, y) \land (\forall y)B(x, y)$;
c. $(\forall x)(\exists y)S(x, y, z) \Rightarrow (\exists x)(\forall y)P(x, y)$.

3.61. For a domain of two elements, construct tables of values of the formulas

a. $(\forall x)[A \lor P(x)] \Leftrightarrow A \lor (\forall x)P(x)$;
b. $(\exists x)[A \land P(x)] \Leftrightarrow A \land (\exists x)P(x)$.

3.62. Using formulas of predicate logic, write propositions expressing the definition of the limit of a function at a point and its negation. (In the second case, reduce the formula to a form such that the quantifiers will not be under the negation symbol.)

3.63. Construct propositions expressing the denial of the following propositions

a. $(\forall x)(\forall y)[x^2 + y^2 > 0]$,

where x and y are variables for real numbers;

b. $(\forall x)(\forall y)[\sqrt{x^2 + y^2} = x + y]$,

where x and y are variables for real numbers;

c. $(\forall x)(\forall y)[\bar{D}(3, x) \land \bar{D}(3, y) \Rightarrow \bar{D}(3, x + y)]$,

where $D(x, y)$ is the predicate "x divides y" and x and y are variables for integers.

3.64 Show that the implications converse to implications 40 and 45 are not universally valid formulas.

* For a proof, cf. Mendelson [6], p. 156, or Church [1], p. 253.

3.65 For what kind of domain is the following rule of inference valid?

$$\frac{(\exists x)P(x)}{(\forall x)P(x)}$$

3.66. What rules of inference can be obtained on the basis of the laws of logic expressed by the universally valid formulas 31–42?

5. TRADITIONAL LOGIC (THE LOGIC OF ONE-PLACE PREDICATES)

Traditional formal logic, the basis of which is constituted by the theory of the so-called "categorical syllogism," which was developed as far back as Aristotle, can be completely translated into the language of the logic of one-place predicates.

By "categorical syllogism," one means in traditional logic an inference in which one draws a conclusion from two premises, where the premises and the conclusion are propositions of any of the following four types:

1. All A are B (the universal affirmative proposition).
2. No A is B (the universal negative proposition).
3. Some A are B (the particular affirmative proposition).
4. Some A are not B (the particular negative proposition).

(A and B are the **terms** of the proposition.)

The premises of the syllogism can be two propositions containing one term in common, for example,

All A are B

and

No B is C.

The problem consists in finding a conclusion in the form of a proposition of one of the four types with the terms A and C.

From the given premises, we get the conclusions

No A is C

(see Figure 13) and

No C is A.

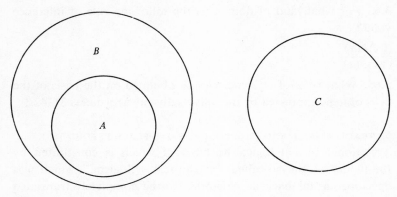

Figure 13

On the other hand, if we take the premises

All A are B

and

Some B are not C,

we cannot draw any conclusion with terms A and C since the given premises do not determine precisely the relationship between the classes (sets) A and C (see Figure 14).

The object of the theory of the categorical syllogism is to find all cases such that from two premises of the types indicated we can draw a conclusion. Of the 256 possible cases, only in 19 does a definite conclusion follow from the premises. These are the so-called *modes* of a categorical syllogism.*

The theory of the categorical syllogism can be constructed in both set-theoretic language (the language of the logic of classes) and the language of the logic of one-place predicates. Here, we shall not construct this theory† but we shall show as examples

---

* In traditional logic, we assume that the truth sets are not empty. If we do not make this assumption, then only 15 of the 19 modes can be justified by means of predicate logic.
† See [3], Chapter 2.

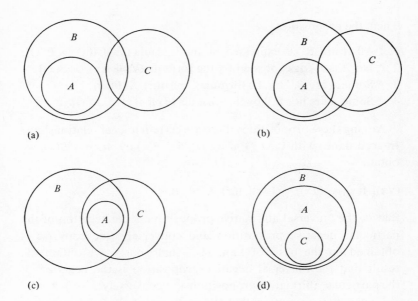

Figure 14

the justification for various modes of a syllogism within the language of predicate logic.

5.1. First of all, let us express the propositions of the four types in the language of the logic of one-place predicates. The terms of the propositions can be regarded as sets (classes) or as properties (one-place predicates). Thus, the proposition "all A are B" can be understood in the sense "all objects belonging to the class A belong to the class B" ("the class A is contained in the class B") or in the sense "all objects possessing property A also possess property B." (Of course, each of these meanings can be reduced to the other if we call membership in the class A property A.) We shall adhere to the second meaning.

We introduce the two one-place predicates:

$A(x)$: x possesses property A,
$B(x)$: x possesses property B.

SEC. 5. TRADITIONAL LOGIC   171

Then, the proposition

1. "All A are B" is expressed by the formula $(\forall x)[A(x) \Rightarrow B(x)]$;
2. "No A is B" is expressed by the formula $(\forall x)[A(x) \Rightarrow \bar{B}(x)]$;
3. "Some A are B," by the formula $(\exists x)[A(x) \wedge B(x)]$;
4. "Some A are not B," by the formula $(\exists x)[A(x) \wedge \bar{B}(x)]$.

Among these propositions there are certain logical relationships. In accordance with Law 33, if we replace $P(x)$ by $A(x) \Rightarrow B(x)$, we obtain

$$(\forall x)[A(x) \Rightarrow B(x)] \Rightarrow \overline{(\exists x)}[A(x) \wedge \bar{B}(x)],$$

that is, the universal affirmative proposition is the negation of the particular negative proposition and conversely (the converse is obtained on the basis of Law 34). Analogously, we obtain the result that the universal negative proposition is the negation of the particular affirmative proposition and conversely.

Under the assumption that the truth set of the predicate $A(x)$ is a nonempty set, that is, that $(\exists x)A(x)$ is a true proposition, we can show with the tools of predicate logic that the universal affirmative proposition implies the particular affirmative proposition and that the universal negative proposition implies the particular negative proposition. (Intuitively, this is perfectly obvious: if it is true that all A are B, then it is true also that some A are B.)

Let us show that the formulas

$$(\forall x)[A(x) \Rightarrow B(x)] \Rightarrow (\exists x)[A(x) \wedge B(x)]$$

and

$$(\forall x)[A(x) \Rightarrow \bar{B}(x)] \Rightarrow (\exists x)[A(x) \wedge \bar{B}(x)]$$

assume the value T for all A and B under the assumption that there exist objects possessing property A, that is, that $(\exists x)A(x)$ is a true proposition.

The formula

$$(\exists x)A(x) \Rightarrow ((\forall x)[A(x) \Rightarrow B(x)] \Rightarrow (\exists x)[A(x) \wedge B(x)])$$

is universally valid. In fact, transforming this formula, we obtain

$(\overline{\exists x})A(x) \lor (\overline{\forall x})[A(x) \Rightarrow B(x)] \lor (\exists x)[A(x) \land B(x)]$ equiv
$(\overline{\exists x})A(x) \lor (\exists x)[A(x) \land \bar{B}(x)] \lor (\exists x)[A(x) \land B(x)]$ equiv
$(\overline{\exists x})A(x) \lor (\exists x)A(x)$ equiv T.

From the true formulas

$(\exists x)A(x) \Rightarrow ((\forall x)[A(x) \Rightarrow B(x)] \Rightarrow (\exists x)[A(x) \land B(x)])$

and

$(\exists x)A(x),$

we obtain, in accordance with the detachment rule (*modus ponens*), the result that the formula

$(\forall x)[A(x) \Rightarrow B(x)] \Rightarrow (\exists x)[A(x) \land B(x)]$

is also a true formula.

If we replace $B(x)$ everywhere by $\bar{B}(x)$, we obtain the result that
$(\forall x)[A(x) \Rightarrow \bar{B}(x)] \Rightarrow (\exists x)[A(x) \land \bar{B}(x)]$
assumes the value T for all A and B (under the condition indicated above).

5.2. Let us look at some specific lines of reasoning that have the form of a categorical syllogism and let us establish their validity with the tools of predicate logic.

a. Consider the following reasoning:

all rational numbers are real numbers,
all integers are rational numbers;

consequently, all integers are real numbers.

This is a syllogism in which both of the premises and the conclusion are universal affirmative propositions. Its validity is easily established with the aid of the syllogism law (29) and the laws expressed by formulas 35 and 42. Let us prove this. Suppose that $A(x)$, $B(x)$, and $C(x)$ are respectively the predicates

x is an integer,
x is a rational number,
x is a real number.

Then, our syllogism can be expressed in the form of the following rule of inference:

$$\frac{(\forall x)[B(x) \Rightarrow C(x)], (\forall x)[A(x) \Rightarrow B(x)]}{(\forall x)[A(x) \Rightarrow C(x)]}.$$

To show the admissibility of this rule of inference, we need only prove the universal validity of the following formula:

$$(\forall x)[B(x) \Rightarrow C(x)] \wedge (\forall x)[A(x) \Rightarrow B(x)] \Rightarrow (\forall x)[A(x) \Rightarrow C(x)].$$

We begin with the universally valid Formula 29:

$$(Y \Rightarrow Z) \wedge (X \Rightarrow Y) \Rightarrow (X \Rightarrow Z).$$

If we replace X by $A(x)$ in this formula, Y by $B(x)$, and Z by $C(x)$, we obtain the universally valid formula

$$(B(x) \Rightarrow C(x)) \wedge (A(x) \Rightarrow B(x)) \Rightarrow (A(x) \Rightarrow C(x)).$$

Since this formula is universally valid, that is, since it assumes the value T for arbitrary values of the predicate variables A, B, and C and the free individual variable x, it follows that the formula obtained from it by binding the individual variable with the universal quantifier is also universally valid; that is,

$$\vDash (\forall x)[(B(x) \Rightarrow C(x)) \wedge (A(x) \Rightarrow B(x)) \Rightarrow (A(x) \Rightarrow C(x))].$$

By applying Formula 42, we obtain

$$\vDash (\forall x)[(B(x) \Rightarrow C(x)) \wedge (A(x) \Rightarrow B(x))] \Rightarrow (\forall x)[A(x) \Rightarrow C(x)]$$

and, finally, by applying Formula 35, we obtain

$$\vDash (\forall x)[B(x) \Rightarrow C(x)] \wedge (\forall x)[A(x) \Rightarrow B(x)] \Rightarrow (\forall x)[A(x) \Rightarrow C(x)]. \tag{1}$$

We have obtained one of the modes of a syllogism. In traditional logic, this mode is known by the name *Barbara*. (The three occurrences of the vowel *a* in this name mean that the premises and conclusion are universal affirmative propositions, denoted by the letter *a*, the initial letter of the latin word *affirmo*, meaning "I assert.")

Some other modes of a syllogism can be reduced to Form 1.

b. Consider the reasoning

no real number is imaginary,
all rational numbers are real numbers;
___
consequently, no rational number is imaginary.

Let us show the validity of this reasoning, which has the form of a syllogism.

We denote the predicate "x is a rational number" by $A(x)$, the predicate "x is a real number" by $B(x)$, and the predicate "x is an imaginary number" by $C(x)$. Then, our syllogism can be written as the following rule of inference:

$$\frac{(\forall x)[B(x) \Rightarrow \overline{C}(x)], \, (\forall x)[A(x) \Rightarrow B(x)]}{(\forall x)[A(x) \Rightarrow \overline{C}(x)]}.$$

One can easily see that this rule of inference reduces to Form 1. Specifically, if we replace $C(x)$ in Form 1 by $\overline{C}(x)$, we obtain

$$\vdash (\forall x)[B(x) \Rightarrow \overline{C}(x)] \wedge (\forall x)[A(x) \Rightarrow B(x)] \Rightarrow (\forall x)[A(x) \Rightarrow \overline{C}(x)].$$

We have obtained the mode *Celarent*. (The vowel sequence *eae* in this word means that the first premise is a universal negative proposition [*e* is the first vowel of the latin word *nego*, meaning "I deny"], the second is a universal affirmative proposition, and the conclusion is a universal negative one.)

c. Consider the reasoning

all integers are rational numbers,
some real numbers are integers;
___
consequently, some real numbers are rational.

We use the following notation:

$A(x)$: x is a real number,
$B(x)$: x is an integer,
$C(x)$: x is a rational number.

Then our syllogism can be written in the form of the following rule of inference

$$\frac{(\forall x)[B(x) \Rightarrow C(x)], \ (\exists x)[A(x) \wedge B(x)]}{(\exists x)[A(x) \wedge C(x)]}.$$

Proving the validity of this mode of the syllogism means proving the universal validity of the formula

$$(\forall x)[B(x) \Rightarrow C(x)] \wedge (\exists x)[A(x) \wedge B(x)] \Rightarrow (\exists x)[A(x) \wedge C(x)]. \quad (2)$$

This can be done, just as in Case 1, on the basis of the following tautology of propositional logic:

$$(Y \Rightarrow Z) \wedge X \wedge Y \Rightarrow X \wedge Z.$$

(The fact that this formula is a tautology is easily shown with the aid of transformations or a truth table.)

The universally valid Formula 2 is called the mode *Darii*. (The vowel sequence *aii* in this word means that the first premise is a universal affirmative proposition, and the second premise and conclusion are particular affirmative propositions, denoted by the second vowel, namely *i*, of the word *affirmo*.)

d. If we substitute $\bar{C}(x)$ for $C(x)$ in Formula 2 (the universal validity of which has already been established), we obtain

$$(\forall x)[B(x) \Rightarrow \bar{C}(x)] \wedge (\exists x)[A(x) \wedge B(x)] \Rightarrow (\exists x)[A(x) \wedge \bar{C}(x)],$$

or the rule of inference

$$\frac{(\forall x)[B(x) \Rightarrow \bar{C}(x)], \ (\exists x)[A(x) \wedge B(x)]}{(\exists x)[A(x) \wedge \bar{C}(x)]}.$$

We have obtained the mode *Ferio*. (The vowel sequence *eio* means that the first premise is a universal negative proposition, that the second one is a particular affirmative one, and that the conclusion is a particular negative one, indicated by the second vowel, namely *o*, of the word *nego*.)

An example of a line of reasoning that has the structure of this mode is the following:

no real number is imaginary,
some complex numbers are real;
---
consequently, some complex numbers are not imaginary.

Fifteen of the nineteen modes of a syllogism reduce to formulas 1 and 2. The remaining four which lead from two universal premises to a particular conclusion require the additional condition that objects possessing property B exist; that is, they require the truth of $(\exists x)B(x)$, where B is the common term of the two premises.

The fact that traditional logic constitutes only a small fragment of predicate logic, one of the logical systems that is a part of mathematical logic, illustrates the effectiveness of the application of mathematical methods to logic.

EXERCISES

3.67. Show that the following lines of reasoning have the structure of modes of a syllogism that can be reduced to formulas 1 and 2.

a. All squares are rhombuses,
   some rectangles are not rhombuses;
   ---
   consequently, some rectangles are not squares.

   (Baroko)

b. No imaginary number is a real number,
   some complex numbers are real numbers;
   ---
   consequently, some complex numbers are not imaginary.

   (Festino)

c. No imaginary number is a real number,
   all rational numbers are real numbers;
   ---
   consequently, no rational number is imaginary.

   (Cesare)

d. All squares are regular polygons,
   no trapezoid is a regular polygon;
   ---
   consequently, no trapezoid is a square

   (Camestres)

3.68. Prove the invalidity of the following lines of reasoning (by using predicate logic).

a. All integers are rational numbers,
some fractions are not integers;

   consequently, some fractions are not rational numbers.

b. All rhombuses are parallelograms,
all rectangles are parallelograms;

   consequently, all rectangles are rhombuses.

c. Some real numbers are rational numbers,
some rational numbers are not integers;

   consequently, some real numbers are not integers.

d. No trapezoid is a regular polygon,
no triangle is a trapezoid;

   consequently, no triangle is a regular polygon.

6. PREDICATE LOGIC WITH EQUALITY. AXIOMATIC CONSTRUCTION OF MATHEMATICAL THEORIES IN THE LANGUAGE OF PREDICATE LOGIC WITH EQUALITY

6.1. As we have already seen, the language of predicate logic is considerably more "expressive" than the language of propositional logic. On the basis of the language of predicate logic, we can construct certain mathematical theories. In the construction of these theories, one uses the equality (coincidence, identity) relationship between objects. This relation is not specific to each of these theories. Therefore, it is frequently used not to refer to special mathematical relationships but to general logical ones.

For example, the equality predicate is necessary for precise expression both of the geometrical proposition

For any two distinct points, there exists no more than a single straight line incident to both

and of the arithmetic proposition

For any two natural numbers, there exists no more than one natural number equal to their sum.

The equality predicate, which we denote by writing $x = y$, assumes the value T if in place of x and y we put the names of one and the same object, and it assumes the value F if we substitute the names of different objects.

This concept of equality (or identity) is characterized by the following conditions: (1) The object x is identical with itself (the condition of reflexivity); and (2) If the object x is identical with the object y, and if x has a certain property A, then y has this same property; that is, coincidence of objects means that they have the same properties.

These conditions determining the equality predicate are expressed by the following formulas:

$$x = x \quad \text{(reflexivity)}; \tag{1}$$
$$(x = y) \Rightarrow (A(x) \Rightarrow A(y)). \tag{2}$$

From formulas 1 and 2 are derived the properties of symmetry and transitivity of the equality predicate.

To carry out this derivation, we use one of the rules of inference that are applied in predicate logic, namely, the rule for replacement of a free individual variable: if in a true formula φ any free individual variable is replaced everywhere it occurs in that formula by another free individual variable (that does not originally appear in the formula), the result is a true formula. We denote by $S_x^y \varphi$ the result of replacement of a free individual variable x by a variable y in the formula φ.

Let $A(x)$ express the following property of the object x: "$x = z$." Then, from Formula 2, we obtain

$$(x = y) \Rightarrow ((x = z) \Rightarrow (y = z)). \tag{3}$$

Let us now show that the truth of $x = y$ implies the truth of $y = x$, that is, let us show that the implication

$$(x = y) \Rightarrow (y = x)$$

expressing the symmetry of the equality predicate is true.

Let us suppose that $x = y$ has the value T. Then, from $x = y$ and Formula 3 we obtain in accordance with the detachment rule

$(x = z) \Rightarrow (y = z)$, (4)

$S_z^z(4) : (x = x) \Rightarrow (y = x)$. (5)

From formulas 1 and 5, we obtain $y = x$ in accordance with the detachment rule.

We have shown that from $x = y$, it follows that $y = x$, that is, we have proved the implication*

$(x = y) \Rightarrow (y = x)$. (6)

One can also easily show the truth of the opposite implication:

$(y = x) \Rightarrow (x = y)$. (7)

$S_x^z(6) : (z = y) \Rightarrow (y = z)$, (8)
$S_y^z(8) : (z = x) \Rightarrow (x = z)$, (9)
$S_z^y(9) : (y = x) \Rightarrow (x = y)$. (10)

From formulas 6 and 7, we get the equivalence of the formulas

$x = y$ and $y = x$.

If in Formula 3 we replace the formula $x = y$ by the formula $y = x$ equivalent to it, we obtain

$(y = x) \Rightarrow ((x = z) \Rightarrow (y = z))$,

or $(y = x) \land (x = z) \Rightarrow (y = z)$ (transitivity of the equality predicate).

Predicate logic, including the equality predicate (with axioms 1 and 2 characterizing this predicate), is also called **predicate logic with equality**.

Let us now express in the language of predicate logic with equality the two propositions introduced at the beginning of this section.

---

* In going from $(x = y) \vdash (y = x)$ to $\vdash (x = y) \Leftrightarrow (y = x)$, we use the deduction theorem (cf. pp. 112–113), which remains valid (under certain conditions) for the predicate calculus.

a. We formulate the proposition

For any two distinct points, there exists no more than one straight line incident to both

as follows:

For any two points A and B and any two straight lines a and b, if these points are distinct (do not coincide) and if the straight lines are incident to them, then these straight lines coincide.

(This simply means that there does not exist more than one straight line incident to these points.)

We denote by $I(A, a)$ the predicate

The point A is incident to the straight line a

or

The straight line a is incident to the point A.

Then, the proposition that we are considering can be written in the form of the following formula:

$$(\forall A)(\forall B)(\forall a)(\forall b)[\overline{A = B} \wedge I(A, a) \wedge I(B, a) \wedge I(A, b) \\ \wedge I(B, b) \Rightarrow (a = b)].$$

b. To write in the form of a formula the proposition

For any two natural numbers, there exists no more than one natural number equal to their sum,

we introduce the three-place predicate $M(x, y, z)$ denoting $x + y = z$.

Then, our proposition can be expressed by the following formula:

$$(\forall x)(\forall y)(\forall z)(\forall u)[M(x, y, z) \wedge M(x, y, u) \Rightarrow (z = u)].$$
$\underbrace{\qquad\qquad\qquad\qquad}_{\in N}$

6.2. To construct a mathematical theory on the basis of predicate logic (with equality), this logical system is supplemented with the primitive concepts of the mathematical theory: individual constants and predicates.

The specific nature of these constants and predicates is described by certain propositions, the set of which constitute the system of proper axioms of the theory. These axioms are expressed by formulas of predicate logic which are not universally valid. They are true only for the specific class of domains possessing the common structure described by the given theory, and they are combined with the universally valid formulas of predicate logic for the derivation or proof of the theorems of the theory.

By a *proof* in such a theory, we mean a finite sequence

$A_1, A_2, \ldots, A_n$

of propositions of the theory such that each proposition either is an axiom or can be obtained from the preceding propositions in this sequence according to the rules of inference of the logic (calculus) of predicates.

If there exists at least one such sequence of propositions that terminates in the proposition L, then L is a **theorem** or a **derived proposition** in the theory. (Of course, every axiom in the theory is a theorem, the proof of which consists of that axiom itself.) Thus, we can introduce into the system of predicate logic characterizations of Euclidean geometry, arithmetic, group theory, etc. and obtain formalizations of the corresponding mathematical theories in the language of that logic.

Let us give two examples. (We confine ourselves to expressing the axioms of the theory and proving one or two theorems in predicate logic with equality).

a. As our first example, let us take a small fragment of two-dimensional geometry, namely, the theory of incidence.

We define a **plane** as a set of elements of two categories: **points**, which we denote by upper-case letters $A, B, C, \ldots$, and **lines**, which we denote by lower-case letters* $a, b, c, \ldots$.

---

* Instead of introducing two categories of elements (essentially, two categories of individual variables, for points and lines), we can introduce two one-place predicates:

In a plane (that is, in a set of points and lines) one defines a two-place predicate that applies to two elements of different categories, that is, to a point and a straight line, this predicate being known as **incidence**. We denote it by $I(A, a)$ and we read it as follows: "the point A is incident to the line a" or "the line a is incident to the point A." This predicate is characterized by the following four axioms known as the incidence axioms:

$$(\forall A)(\forall B)(\exists a)[I(A, a) \land I(B, a)] \qquad (i.1)$$

(For any two points, there exists a line incident to them);

$$(\forall A)(\forall B)(\forall a)(\forall b)[\overline{A = B} \land I(A, a) \land I(B, a) \land I(A, b) \\ \land I(B, b) \Rightarrow (a = b)] \qquad (i.2)$$

(for any two distinct points, there does not exist more than one line incident to them);

$$(\forall a)(\exists A)(\exists B)[\overline{A = B} \land I(A, a) \land I(B, a)] \qquad (i.3)$$

(for an arbitrary line, there exist two distinct points incident to it);

$$(\exists A)(\exists B)(\exists C)(\forall a)[\overline{I(A, a) \land I(B, a) \land I(C, a)}] \qquad (i.4)$$

(there exist three points not all incident to any one line).

The class of propositions derivable from axioms $i.1$–$i.4$ by means of predicate logic with equality constitutes the theory of incidence (in a plane).

As an example, let us prove the theorem

THEOREM 1
For any two distinct lines, there does not exist more than one point incident to them; that is,

$$(\forall a)(\forall b)(\forall A)(\forall B)[\overline{a = b} \land I(A, a) \land I(B, a) \\ \land I(A, b) \land I(B, b) \Rightarrow (A = B)].$$

$P(x)$: x is a point
and
$L(x)$: x is a straight line.
(Then, instead of the quantifiers with respect to the two kinds of variables, we need only use quantifiers restricted by these predicates.)

Proof:

1. $(\forall A)(\forall B)(\forall a)(\forall b)[\overline{A = B} \land I(A, a) \land I(B, a) \land I(A, b)$
$\land I(B, b) \Rightarrow (a = b)]$ \quad (*i*.2)
2. $(\forall A)(\forall B)(\forall a)(\forall b)[\overline{a = b} \land I(A, a) \land I(B, a)$
$\land I(A, b) \land I(B, b) \Rightarrow \overline{\overline{A = B}}]$
3. $(\forall A)(\forall B)(\forall a)(\forall b)[\overline{a = b} \land I(A, a) \land I(B, a)$
$\land I(A, b) \land I(B, b) \Rightarrow (A = B)]$
4. $(\forall a)(\forall b)(\forall A)(\forall B)[\overline{a = b} \land I(A, a) \land I(B, a)$
$\land I(A, b) \land I(B, b) \Rightarrow (A = B)].$

This proof is a sequence of four propositions. The first is Axiom *i*.2. The second is obtained from the first in accordance with the law of extended contraposition (24). The third is obtained from the second in accordance with the law of double negation (1). The fourth, which is the theorem being proved, is obtained from the third in accordance with Law 37, which tells us that we can switch the positions of quantifiers of the same kind.

(In view of the fact that all the laws that we used in this proof are expressed by formulas which are equivalences, this proof can be reversed; that is, the sequence of propositions 4–3–2–1 constitutes a proof of Proposition *i*.2 if Theorem 1 is taken as an axiom.)

On the basis of the incidence predicate, we can introduce other predicates with the aid of appropriate definitions.

We denote by $a \times b$ the two-place predicate of *intersection* as applied to two lines. It is read as follows: "line a intersects line b." This predicate is defined as follows:

$$a \times b \text{ equiv}_{Df} \overline{a = b} \land (\exists A)[I(A, a) \land I(A, b)]$$

(The statement "a intersects b" means by definition that a and b are distinct lines and that there exists a point incident to both.) Obviously, the notation $a \times b$ serves as an abbreviation for a compound predicate constructed from the elementary predicates of equality and incidence with the aid of the operations of predicate logic.

The two-place predicate of *parallelism* applied to two lines and denoted by $a||b$ (read "the line a is parallel to the line b") is defined as the negation of the predicate of intersection:

$(a||b)$ equiv$_{Df}$ $\overline{a \times b}$

(The statement "a is parallel to b" means by definition that "a does not intersect b.")

From the definition of the predicate of intersection, we obtain

$(a||b)$ equiv $\overline{a = b \wedge (\exists A)[I(A, a) \wedge I(A, b)]}$
    equiv $(a = b) \vee (\overline{\exists A})[I(A, a) \wedge I(A, b)]$

(a is parallel to b if and only if a coincides with b or there exists no point incident to both a and b).*

b. Let us show how one can construct on the basis of the language of predicate logic with equality a formal system of the arithmetic of natural numbers.

As the range of values of the individual variables, we take a nonempty set N, the elements of which are called **natural numbers**. On the set N^2, we define the two-place predicate $R(x, y)$:

y is the immediate successor of x.

If a pair (x, y) is in the relation R, we also write $y = x'$; that is,

$(y = x')$ equiv$_{Df}$ $R(x, y)$.

We introduce also an individual constant, a particular natural number denoted by the symbol 1 (so that $1 \in N$).

The predicate $R(x, y)$ and the constant 1 are characterized by the following axioms:

$(\forall x)\bar{R}(x, 1)$ or $(\forall x)\overline{[1 = x']}$ or $\overline{(\exists x)}[1 = x']$ $\qquad(a1)$

(there does not exist a natural number of which 1 is the immediate successor);

$(\forall x)(\exists y)R(x, y)$ or $(\forall x)(\exists y)[y = x']$ $\qquad(a2)$

* The definition given here of the relation of parallelism differs from the traditional one used in school. It is obviously more convenient for us to make the relation of parallelism reflexive, symmetric, and transitive, that is, an example of what is known in logic as an equivalence relation.

(for every natural number, there exists a natural number that is the immediate successor of the first);

$$(\forall x)(\forall y)(\forall u)(\forall v)[(x = y) \wedge R(x, u) \wedge R(y, v) \Rightarrow (u = v)], \quad (a3)$$

or, in a different way of writing,

$$(\forall x)(\forall y)[(x = y) \Rightarrow (x' = y')]$$

(for every natural number, there exists no more than one natural number that is the immediate successor of the first);

$$(\forall x)(\forall y)(\forall u)(\forall v)[(u = v) \wedge R(x, u) \wedge R(y, v) \Rightarrow (x = y)],$$
$$\text{or} \quad (\forall x)(\forall y)[(x' = y') \Rightarrow (x = y)] \quad (a4)$$

(every natural number is the immediate successor of no more than one natural number);

$$A(1) \wedge (\forall x)(\forall y)[A(x) \wedge R(x, y) \Rightarrow A(y)] \Rightarrow (\forall z)A(z),$$
$$\text{or} \quad A(1) \wedge (\forall x)[A(x) \Rightarrow A(x')] \Rightarrow (\forall z)A(z). \quad (a5)$$

(This is the mathematical induction axiom: if 1 possesses a property A and if, for every natural number, the fact that it possesses property A implies that its immediate successor also possesses property A, then every natural number possesses property A.)

Remark: The induction axiom ($a5$) differs from axioms $a1$–$a4$ in that it contains the "predicate variable" A. Therefore, it is not a proposition but a logical function of this predicate variable, which becomes a true proposition for an arbitrary value of A. Thus, Axiom $a5$ is a *scheme* of axioms from which we can obtain specific axioms (in the proper sense of the word) for different values of the predicate variable A.

As an example, let us prove the theorem

Every natural number is distinct from its immediate successor;

that is,

$(\forall z)[\overline{z = z'}]$.

1. $(\forall x)[\overline{1 = x'}]$ ($a1$);

2. $\overline{1 = 1'}$ (from Step 1 and Law 44* in accordance with *modus ponens*);

3. $(\forall x)(\forall y)[(x' = y') \Rightarrow (x = y)]$ (a4);

4. $(\forall x)(\forall y)[\overline{x = y} \Rightarrow \overline{x' = y'}]$ (from Step 3 in accordance with the contraposition rule);

5. $(\forall x)[\overline{x = x'} \Rightarrow \overline{x' = (x')'}]$ (from Step 4 by replacing y by x');

6. $\overline{1 = 1'} \wedge (\forall x)[\overline{x = x'} \Rightarrow \overline{x' = (x')'}]$ $\left(\text{from Steps 2 and 5 in accordance with the rule } \dfrac{\varphi_1, \varphi_2}{\varphi_1 \wedge \varphi_2}\right)$;

7. $A(1) \wedge (\forall x)[A(x) \Rightarrow A(x')] \Rightarrow (\forall z)A(z)$ (a5);

8. $\overline{1 = 1'} \wedge (\forall x)[\overline{x = x'} \Rightarrow \overline{x' = (x')'}] \Rightarrow (\forall z)[\overline{z = z'}]$ (from Step 7 by replacing $A(x)$ by $\overline{x = x'}$);

9. $(\forall z)[\overline{z = z'}]$ (from Steps 8 and 6 in accordance with the detachment rule *modus ponens*).

Axioms $a1$–$a5$ characterize the set of natural numbers only from the point of view of the relationship of immediate succession.

On the basis of this relation, we can introduce the operations of addition and multiplication as three-place predicates defined on the set N^3 and characterized by special axioms.

Thus, we introduce the three-place predicates $S(x, y, z)$ (which we usually write $x + y = z$) and $P(x, y, z)$ (which we usually write $x \cdot y = z$) characterized by the following axioms

$(\forall x)S(x, 1, x')$;   (a6)
$(\forall x)(\forall y)(\forall z)[S(x, y, z) \Rightarrow S(x, y', z')]$;   ($a$7)
$(\forall x)P(x, 1, x)$;   (a8)
$(\forall x)(\forall y)(\forall z)(\forall u)[P(x, y, z) \wedge S(x, z, u) \Rightarrow P(x, y', u)]$.   ($a$9)

In ordinary notation,

$(\forall x)[x + 1 = x']$;

* Cf. footnote on p. 167.

$(\forall x)(\forall y)[x + y' = (x + y)']$;
$(\forall x)[x \cdot 1 = x]$;
$(\forall x)(\forall y)[x \cdot y' = xy + x]$.

EXERCISES

3.69. Using the notation of Section 6.2a, p. 182, express in the language of predicate logic with equality the axioms of Euclid and Lobachevski on parallels.

3.70. Express with the aid of the three-place predicates $S(x, y, z)$ and $P(x, y, z)$ defined above the commutativity and associativity of addition and multiplication and the distributivity of multiplication with respect to addition.

3.71. With the aid of the predicate S, define the predicate "$x < y$" (that is, "x is less than y").

3.72. Write formulas expressing the antisymmetry, antireflexivity, and transitivity of the predicate $x < y$. If this predicate is taken as primitive, how can we express the predicate $R(x, y)$ in terms of it?

3.73. Write in the language of predicate logic with equality the fact that the relation $<$ orders the set of natural numbers.

APPENDIXES

I. A PROOF OF THE DUALITY PRINCIPLE FOR PROPOSITIONAL LOGIC

We shall restrict ourselves in this Appendix to formulas of propositional logic built up from variables and the letters T and F by means of the connectives $^{-}$, $\wedge$, $\vee$ only. The *dual* φ' of a formula φ is obtained from φ by interchanging $\wedge$ and $\vee$, and interchanging T and F. Clearly, φ'' is identical with φ.

LEMMA

Let φ^b be obtained from φ by replacing all variables by their negations. Then: φ^b equiv $\overline{\varphi'}$.

(Example: If φ is $X \vee (Y \wedge (\overline{X} \vee Z))$, then φ^b is $\overline{X} \vee (\overline{Y} \wedge (\overline{\overline{X}} \vee \overline{Z}))$.)

PROOF

This is a simple consequence of applying equivalences 16–17 (p. 52) to $\overline{\varphi'}$ until negations no longer apply to conjunctions or disjunctions. To make this idea precise, we carry out the proof by induction on the number k of occurrences of connectives in φ.

Case. 1. $k = 0$.

Case 1a. φ is a variable, say X. Then φ^b is $\overline{X}$, φ' is φ, and $\overline{X}$ equiv $\overline{X}$.

Case 1b. φ is T. Then φ^b is T, φ' is F, and T equiv $\overline{F}$.

Case 1c. φ is F. Then φ^b is F, φ' is T, and F equiv $\overline{T}$.

Case. 2. $k > 0$ and the lemma holds for all formulas having fewer than k occurrences of connectives.

Case 2a. φ is $\overline{\psi}$. Then, by hypothesis, ψ^b equiv $\overline{\psi'}$. Hence, $\overline{\psi^b}$ equiv $\overline{\overline{\psi'}}$. By definition, φ^b is $\overline{\psi^b}$ and φ' is $\overline{\psi'}$. Therefore, φ^b equiv $\overline{\varphi'}$.

Case 2b. φ is $\psi \wedge \theta$. Then, by hypothesis, ψ^b equiv $\overline{\psi'}$ and θ^b equiv $\overline{\theta'}$. Now, φ^b is $\psi^b \wedge \theta^b$ and φ' is $\psi' \vee \theta'$. Hence,
φ^b equiv $\psi^b \wedge \theta^b$
  equiv $\overline{\psi'} \wedge \overline{\theta'}$ (by Rule 2, p. 54)
  equiv $\overline{\psi' \vee \theta'}$ (by Equivalence 17, p. 52)
  equiv $\overline{\varphi'}$.

Case 2c. φ is $\psi \vee \theta$. Then, by hypothesis, ψ^b equiv $\overline{\psi'}$ and θ^b equiv $\overline{\theta'}$. Now, φ^b is $\psi^b \vee \theta^b$ and φ' is $\psi' \wedge \theta'$. Hence,

φ^b equiv $\psi^b \vee \theta^b$
  equiv $\overline{\psi'} \vee \overline{\theta'}$ (by Rule 2, p. 54)
  equiv $\overline{\psi' \wedge \theta'}$ (by Equivalence 16, p. 52)
  equiv $\overline{\varphi'}$.

DUALITY PRINCIPLE

If φ equiv ψ, then φ' equiv ψ'.

PROOF

Assume φ equiv ψ. Then, by Rule 1, p. 54, φ^b equiv ψ^b. Hence, by the lemma, $\overline{\varphi'}$ equiv $\overline{\psi'}$, and, therefore, φ' equiv ψ'.

II. A PROOF OF THE DEDUCTION THEOREM FOR THE PROPOSITIONAL CALCULUS

DEDUCTION THEOREM

If $\varphi_1, \ldots, \varphi_{n-1}, \varphi_n \vdash \varphi$, then $\varphi_1, \ldots, \varphi_{n-1} \vdash \varphi_n \Rightarrow \varphi$. (Hence, by iteration of this result, if $\varphi_1, \ldots, \varphi_n \vdash \varphi$, then $\vdash \varphi_1 \Rightarrow (\varphi_2 \Rightarrow (\cdots (\varphi_n \Rightarrow \varphi) \cdots))$.)

PROOF

(1) As a preliminary, we first show that
$\vdash \psi \Rightarrow \psi$
for any formula ψ.

1. $(A \Rightarrow (B \Rightarrow C)) \Rightarrow ((A \Rightarrow B) \Rightarrow (A \Rightarrow C))$ (I.2)
2. $(\psi \Rightarrow ((\psi \Rightarrow \psi) \Rightarrow \psi))$
   $\Rightarrow ((\psi \Rightarrow (\psi \Rightarrow \psi)) \Rightarrow (\psi \Rightarrow \psi))$ ($\prod_{A,B,C}^{\psi,\psi \Rightarrow \psi, \psi}$ (1))
3. $A \Rightarrow (B \Rightarrow A)$ (I.1)
4. $\psi \Rightarrow ((\psi \Rightarrow \psi) \Rightarrow \psi)$ ($\prod_{A,B}^{\psi, \psi \Rightarrow \psi}$ (3))
5. $(\psi \Rightarrow (\psi \Rightarrow \psi)) \Rightarrow (\psi \Rightarrow \psi)$ (2, 4, DR)
6. $\psi \Rightarrow (\psi \Rightarrow \psi)$ ($\prod_{A,B}^{\psi,\psi}$ (3))
7. $\psi \Rightarrow \psi$ (5, 6, DR)

(2) Now we assume $\varphi_1, \ldots, \varphi_{n-1}, \varphi_n \vdash \varphi$. We must show that $\varphi_1, \ldots, \varphi_{n-1} \vdash \varphi_n \Rightarrow \varphi$. Let $\psi_1, \ldots, \psi_k$ be a derivation of φ from $\varphi_1, \ldots, \varphi_{n-1}, \varphi_n$. Thus, ψ_k is φ, and, for each ψ_i, either ψ_i is a substitution instance of an axiom, or ψ_i is one of the hypotheses $\varphi_1, \ldots, \varphi_{n-1}, \varphi_n$, or ψ_i follows from preceding formulas in the

sequence (that is, from some of $\psi_1, \ldots, \psi_{i-1}$) by the detachment rule. Let us now show that $\varphi_1, \ldots, \varphi_{n-1} \vdash \varphi_n \Rightarrow \psi_i$, for each ψ_i. (The desired result is just the special case when $i = k$, since ψ_k is φ.) We prove this by mathematical induction with respect to i. Thus, our inductive hypothesis is that $\varphi_1, \ldots, \varphi_{n-1} \vdash \varphi_n \Rightarrow \psi_j$ holds for every $j < i$, and we must then show that $\varphi_1, \ldots, \varphi_{n-1} \vdash \varphi_n \Rightarrow \psi_i$. (When $i = 1$, our inductive hypothesis is vacuous and the proof that we shall give simply shows that $\varphi_1, \ldots, \varphi_{n-1} \vdash \varphi_n \Rightarrow \psi_1$.)

Case i. ψ_i is a substitution instance of an axiom. Hence, $\vdash \psi_i$. But, $\vdash \psi_i \Rightarrow (\varphi_n \Rightarrow \psi_i)$, by substitution in Axiom I.1. Hence, by the detachment rule, $\vdash \varphi_n \Rightarrow \psi_i$, and, all the more strongly, $\varphi_1, \ldots, \varphi_{n-1} \vdash \varphi_n \Rightarrow \psi_i$.

Case ii. ψ_i is one of the premises $\varphi_1, \ldots, \varphi_{n-1}$, say φ_r. Then the sequence

$\varphi_r \Rightarrow (\varphi_n \Rightarrow \varphi_r)$ (substitution instance of Axiom I.1)
φ_r (Hypothesis)
$\varphi_n \Rightarrow \varphi_r$ (DR)

shows that $\varphi_r \vdash \varphi_n \Rightarrow \varphi_r$, that is, $\varphi_r \vdash \varphi_n \Rightarrow \psi_i$. Hence, all the more strongly, $\varphi_1, \ldots, \varphi_{n-1} \vdash \varphi_n \Rightarrow \psi_i$.

Case iii. ψ_i is the premise φ_n. By the preliminary observation (1), $\vdash \varphi_n \Rightarrow \varphi_n$, that is, $\vdash \varphi_n \Rightarrow \psi_i$. All the more strongly, $\varphi_1, \ldots, \varphi_{n-1} \vdash \varphi_n \Rightarrow \psi_i$.

Case iv. ψ_i follows by the detachment rule from two preceding formulas ψ_s and ψ_t. Hence, ψ_s must be of the form $\psi_t \Rightarrow \psi_i$ (or similarly, ψ_t must be $\psi_s \Rightarrow \psi_i$). Now, by our inductive hypothesis,

$\varphi_1, \ldots, \varphi_{n-1} \vdash \varphi_n \Rightarrow \psi_t,$ (*)

and

$\varphi_1, \ldots, \varphi_{n-1} \vdash \varphi_n \Rightarrow \underbrace{(\psi_t \Rightarrow \psi_i)}_{\psi_s}.$ (**)

Now, by using the substitution instance
$(\varphi_n \Rightarrow (\psi_t \Rightarrow \psi_i)) \Rightarrow ((\varphi_n \Rightarrow \psi_t) \Rightarrow (\varphi_n \Rightarrow \psi_i))$
of Axiom I.2, together with (**) and the detachment rule, we obtain

$\varphi_1, \ldots, \varphi_{n-1} \vdash (\varphi_n \Rightarrow \psi_i) \Rightarrow (\varphi_n \Rightarrow \psi_i)$,
and this, together with (*) and the detachment rule, yields
$\varphi_1, \ldots, \varphi_{n-1} \vdash \varphi_n \Rightarrow \psi_i$.
This completes the inductive proof.

III. A PROOF OF THE COMPLETENESS THEOREM FOR THE PROPOSITIONAL CALCULUS

LEMMA 1

For any formulas φ and ψ,
- (a) $\vdash \bar{\varphi} \Rightarrow \overline{\varphi \wedge \psi}$
- (b) $\vdash \bar{\psi} \Rightarrow \overline{\varphi \wedge \psi}$
- (c) $\varphi, \psi \vdash \varphi \wedge \psi$
- (d) $\bar{\varphi} \vdash \varphi \Rightarrow \psi$ (Hence, $\vdash \bar{\varphi} \Rightarrow (\varphi \Rightarrow \psi)$, by the deduction theorem.)
- (e) $\bar{\varphi}, \bar{\psi} \vdash \overline{\varphi \vee \psi}$
- (f) $\varphi, \bar{\psi} \vdash \overline{\varphi \Rightarrow \psi}$
- (g) $\bar{\varphi} \Rightarrow \bar{\psi} \vdash \psi \Rightarrow \varphi$
- (h) $\varphi \Rightarrow \psi, \bar{\varphi} \Rightarrow \psi \vdash \psi$

PROOF

(a)
1. $\varphi \wedge \psi \Rightarrow \varphi$ — $\prod_{A,B}^{\varphi,\psi}$ (II.1)
2. $(\varphi \wedge \psi \Rightarrow \varphi) \Rightarrow (\bar{\varphi} \Rightarrow \overline{\varphi \wedge \psi})$ — $\prod_{A,B}^{\varphi \wedge \psi, \varphi}$ (IV.1)
3. $\bar{\varphi} \Rightarrow \overline{\varphi \wedge \psi}$ — 1, 2, DR

(b)
1. $\varphi \wedge \psi \Rightarrow \psi$ — $\prod_{A,B}^{\varphi,\psi}$ (II.2)
2. $(\varphi \wedge \psi \Rightarrow \psi) \Rightarrow (\bar{\psi} \Rightarrow \overline{\varphi \wedge \psi})$ — $\prod_{A,B}^{\varphi \wedge \psi, \psi}$ (IV.1)
3. $\bar{\psi} \Rightarrow \overline{\varphi \wedge \psi}$ — 1, 2, DR

(c)
1. φ — Hypothesis
2. ψ — Hypothesis
3. $\varphi \Rightarrow ((\varphi \Rightarrow (\varphi \Rightarrow \varphi)) \Rightarrow \varphi)$ — $\prod_{A,B}^{\varphi, \varphi \Rightarrow (\varphi \Rightarrow \varphi)}$ (I.1)
4. $(\varphi \Rightarrow (\varphi \Rightarrow \varphi)) \Rightarrow \varphi$ — 1, 3, DR
5. $(4) \Rightarrow [((\varphi \Rightarrow (\varphi \Rightarrow \varphi)) \Rightarrow \psi)$
   $\Rightarrow ((\varphi \Rightarrow (\varphi \Rightarrow \varphi)) \Rightarrow \varphi \wedge \psi)]$ — $\prod_{A,B,C}^{\varphi \Rightarrow (\varphi \Rightarrow \varphi), \varphi, \psi}$ (II.3)
6. $((\varphi \Rightarrow (\varphi \Rightarrow \varphi)) \Rightarrow \psi)$
   $\Rightarrow ((\varphi \Rightarrow (\varphi \Rightarrow \varphi)) \Rightarrow \varphi \wedge \psi)$ — 4, 5, DR
7. $\psi \Rightarrow ((\varphi \Rightarrow (\varphi \Rightarrow \varphi)) \Rightarrow \psi)$ — $\prod_{A,B}^{\psi, \varphi \Rightarrow (\varphi \Rightarrow \varphi)}$ (I.1)
8. $(\varphi \Rightarrow (\varphi \Rightarrow \varphi)) \Rightarrow \psi$ — 2, 7, DR
9. $(\varphi \Rightarrow (\varphi \Rightarrow \varphi)) \Rightarrow \varphi \wedge \psi$ — 6, 8, DR

		10.	$\varphi \Rightarrow (\varphi \Rightarrow \varphi)$	$\prod_{A,B}^{\varphi,\varphi}$ (I.1)
		11.	$\varphi \wedge \psi$	9, 10, DR
(d)		1.	$\bar{\varphi}$	Hypothesis
		2.	φ	Hypothesis
		3.	$\bar{\varphi} \Rightarrow (\bar{\psi} \Rightarrow \bar{\varphi})$	$\prod_{A,B}^{\bar{\varphi},\bar{\psi}}$ (I.1)
		4.	$\bar{\psi} \Rightarrow \bar{\varphi}$	1, 3, DR
		5.	$(\bar{\psi} \Rightarrow \bar{\varphi}) \Rightarrow (\bar{\bar{\varphi}} \Rightarrow \bar{\bar{\psi}})$	$\prod_{A,B}^{\bar{\psi},\bar{\varphi}}$ (IV.1)
		6.	$\bar{\bar{\varphi}} \Rightarrow \bar{\bar{\psi}}$	4, 5, DR
		7.	$\varphi \Rightarrow \bar{\bar{\varphi}}$	$\prod_{A}^{\varphi}$ (IV.2)
		8.	$\bar{\bar{\varphi}}$	2, 7, DR
		9.	$\bar{\bar{\psi}}$	6, 8, DR
		10.	$\bar{\bar{\psi}} \Rightarrow \psi$	$\prod_{A}^{\psi}$ (IV.3)
		11.	ψ	9, 10, DR

Thus, we have: $\bar{\varphi}, \varphi \vdash \psi$.
By the deduction theorem (Appendix II), $\bar{\varphi} \vdash \varphi \Rightarrow \psi$.

(e)		1.	$\bar{\varphi}$	Hypothesis
		2.	$\bar{\psi}$	Hypothesis
		3.	$(\varphi \Rightarrow \psi) \Rightarrow ((\psi \Rightarrow \psi)$ $\Rightarrow (\varphi \vee \psi) \Rightarrow \psi))$	$\prod_{A,B,C}^{\varphi,\psi,\psi}$ (III.3)
		4.	$\bar{\varphi} \Rightarrow (\varphi \Rightarrow \psi)$	Part (d) above
		5.	$\varphi \Rightarrow \psi$	1, 4, DR
		6.	$(\psi \Rightarrow \psi) \Rightarrow ((\varphi \vee \psi) \Rightarrow \psi)$	3, 5, DR
		7.	$\psi \Rightarrow \psi$	Appendix II, (1)
		8.	$(\varphi \vee \psi) \Rightarrow \psi$	6, 7, DR
		9.	$((\varphi \vee \psi) \Rightarrow \psi) \Rightarrow (\bar{\psi} \Rightarrow \overline{\varphi \vee \psi})$	$\prod_{A,B}^{\varphi \vee \psi, \psi}$ (IV.1)
		10.	$\bar{\psi} \Rightarrow \overline{\varphi \vee \psi}$	8, 9, DR
		11.	$\overline{\varphi \vee \psi}$	2, 10, DR

(f) Clearly, by the detachment rule, $\varphi, \varphi \Rightarrow \psi \vdash \psi$.
Hence, by the deduction theorem (Appendix II), $\varphi \vdash (\varphi \Rightarrow \psi) \Rightarrow \psi$.
But $\vdash ((\varphi \Rightarrow \psi) \Rightarrow \psi) \Rightarrow (\bar{\psi} \Rightarrow \overline{\varphi \Rightarrow \psi})$ by $\prod_{A,B}^{\varphi \Rightarrow \psi, \psi}$ (IV.1).
Hence, $\varphi \vdash \bar{\psi} \Rightarrow \overline{\varphi \Rightarrow \psi}$.
Therefore, $\varphi, \bar{\psi} \vdash \overline{\varphi \Rightarrow \psi}$ by the detachment rule.

(g)		1.	$\bar{\varphi} \Rightarrow \bar{\psi}$	Hypothesis
		2.	ψ	Hypothesis
		3.	$(\bar{\varphi} \Rightarrow \bar{\psi}) \Rightarrow (\bar{\bar{\psi}} \Rightarrow \bar{\bar{\varphi}})$	$\prod_{A,B}^{\bar{\varphi},\bar{\psi}}$ (IV.1)

4. $\bar{\bar{\bar{\psi}}} \Rightarrow \varphi$	1, 3, DR
5. $\psi \Rightarrow \bar{\bar{\bar{\psi}}}$	$\prod_{A}^{\psi}$ (IV.2)
6. $\bar{\bar{\bar{\psi}}}$	2, 5, DR
7. $\bar{\bar{\bar{\varphi}}}$	4, 6, DR
8. $\bar{\bar{\bar{\varphi}}} \Rightarrow \varphi$	$\prod_{A}^{\varphi}$ (IV.3)
9. φ	7, 8, DR

Thus, $\bar{\varphi} \Rightarrow \bar{\psi}, \psi \vdash \varphi$.

Hence, by the deduction theorem (Appendix II), $\bar{\varphi} \Rightarrow \bar{\psi} \vdash \psi \Rightarrow \varphi$.

(h)
1. $\varphi \Rightarrow \psi$	Hypothesis
2. $\bar{\psi}$	Hypothesis
3. $(\varphi \Rightarrow \psi) \Rightarrow (\bar{\psi} \Rightarrow \bar{\varphi})$	$\prod_{A,B}^{\varphi,\psi}$ (IV.1)
4. $\bar{\psi} \Rightarrow \bar{\varphi}$	1, 3, DR
5. $\bar{\varphi}$	2, 4, DR
6. $\overline{\bar{\varphi} \Rightarrow \psi}$	2, 5, Part (f) above

Thus, $\varphi \Rightarrow \psi, \bar{\psi} \vdash \overline{\bar{\varphi} \Rightarrow \psi}$.

Hence, by the deduction theorem (Appendix II), $\varphi \Rightarrow \psi \vdash \bar{\psi} \Rightarrow \overline{\bar{\varphi} \Rightarrow \psi}$.

Therefore, by Part (g), $\varphi \Rightarrow \psi \vdash (\bar{\varphi} \Rightarrow \psi) \Rightarrow \psi$.

By the detachment rule, $\varphi \Rightarrow \psi, \bar{\varphi} \Rightarrow \psi \vdash \psi$.

LEMMA 2

Let φ be a formula whose variables occur in the list $\mathscr{B}_1, \ldots, \mathscr{B}_n$. Given any assignment of truth values to $\mathscr{B}_1, \ldots, \mathscr{B}_n$. This determines a truth value for φ. Let

$$\mathscr{B}'_i = \begin{cases} \mathscr{B}_i & \text{if } \mathscr{B}_i \text{ is T} \\ \overline{\mathscr{B}_i} & \text{if } \mathscr{B}_i \text{ is F} \end{cases}$$

and

$$\varphi' = \begin{cases} \varphi & \text{if } \varphi \text{ is T} \\ \bar{\varphi} & \text{if } \varphi \text{ is F} \end{cases}$$

Then
$$\mathscr{B}'_1, \ldots, \mathscr{B}'_n \vdash \varphi'.$$

PROOF

We shall prove this result by induction on the number k of occurrences of connectives in φ.

Case 1. $k = 0$. Then φ is $\mathscr{B}_i$ for some i. But, $\mathscr{B}_i \vdash \mathscr{B}_i$ and $\overline{\mathscr{B}_i} \vdash \overline{\mathscr{B}_i}$. Hence, $\mathscr{B}'_i \vdash \varphi'$, and, all the more strongly, $\mathscr{B}'_1, \ldots, \mathscr{B}'_n \vdash \varphi'$.

Case. 2. $k > 0$. We assume the inductive hypothesis that the lemma holds for all formulas with fewer than k occurrences of connectives.

Case 2a. φ is $\overline{\psi}$.

If φ is T, then ψ is F, φ' is φ, and ψ' is $\overline{\psi}$. Hence, by inductive hypothesis, $\mathscr{B}'_1, \ldots, \mathscr{B}'_n \vdash \overline{\psi}$. But φ' is $\overline{\psi}$.

If φ is F, then ψ is T, φ' is $\overline{\varphi}$, and ψ' is ψ. Hence, by inductive hypothesis, $\mathscr{B}'_1, \ldots, \mathscr{B}'_n \vdash \psi$. By Axiom IV.2 and DR, $\mathscr{B}'_1, \ldots, \mathscr{B}'_n \vdash \overline{\overline{\psi}}$. But φ' is $\overline{\overline{\psi}}$.

Case 2b. φ is $\psi \wedge \theta$.

If ψ is F, then φ is F, ψ' is $\overline{\psi}$, and φ' is $\overline{\varphi}$. By inductive hypothesis, $\mathscr{B}'_1, \ldots, \mathscr{B}'_n \vdash \overline{\psi}$. By Lemma 1(a) above, and DR, $\mathscr{B}'_1, \ldots, \mathscr{B}'_n \vdash \overline{\psi \wedge \theta}$. But φ' is $\overline{\psi \wedge \theta}$.

If θ is F, then φ is F, φ' is $\overline{\varphi}$, and θ' is $\overline{\theta}$. By inductive hypothesis, $\mathscr{B}'_1, \ldots, \mathscr{B}'_n \vdash \overline{\theta}$. By Lemma 1(b) above and DR, $\mathscr{B}'_1, \ldots, \mathscr{B}'_n \vdash \overline{\psi \wedge \theta}$. But φ' is $\overline{\psi \wedge \theta}$.

If ψ is T and θ is T, then φ is T, φ' is φ, ψ' is ψ, and θ' is θ. By inductive hypothesis, $\mathscr{B}'_1, \ldots, \mathscr{B}'_n \vdash \psi$, and $\mathscr{B}'_1, \ldots, \mathscr{B}'_n \vdash \theta$. By Lemma 1(c) above and DR, $\mathscr{B}'_1, \ldots, \mathscr{B}'_n \vdash \psi \wedge \theta$. But φ' is $\psi \wedge \theta$.

Case 2c. φ is $\psi \vee \theta$.

If ψ is T, then φ is T, ψ' is ψ, and φ' is φ. By inductive hypothesis, $\mathscr{B}'_1, \ldots, \mathscr{B}'_n \vdash \psi$. By $\prod_{A,B}^{\psi,\theta}$ (III.1) and DR, $\mathscr{B}'_1, \ldots, \mathscr{B}'_n \vdash \psi \vee \theta$. But φ' is $\psi \vee \theta$.

If θ is T, then φ is T, θ' is θ, and φ' is φ. By inductive hypothesis, $\mathscr{B}'_1, \ldots, \mathscr{B}'_n \vdash \theta$. By $\prod_{A,B}^{\psi,\theta}$ (III.2) and DR, $\mathscr{B}'_1, \ldots, \mathscr{B}'_n \vdash \psi \vee \theta$. But φ' is $\psi \vee \theta$.

If ψ is F and θ is F, then φ is F, φ' is $\overline{\varphi}$, ψ' is $\overline{\psi}$, and θ' is $\overline{\theta}$. By inductive hypothesis, $\mathscr{B}'_1, \ldots, \mathscr{B}'_n \vdash \overline{\psi}$, and $\mathscr{B}'_1, \ldots, \mathscr{B}'_n \vdash \overline{\theta}$. By Lemma 1(e) above, $\mathscr{B}'_1, \ldots, \mathscr{B}'_n \vdash \overline{\psi \vee \theta}$. But φ' is $\overline{\psi \vee \theta}$.

Case 2d. φ is $\psi \Rightarrow \theta$.

If ψ is F, then φ is T, φ' is φ, and ψ' is $\overline{\psi}$. Then, by inductive hypothesis, $\mathscr{B}'_1, \ldots, \mathscr{B}'_n \vdash \overline{\psi}$. By Lemma 1(d), $\mathscr{B}'_1, \ldots, \mathscr{B}'_n \vdash \psi \Rightarrow \theta$. But φ' is $\psi \Rightarrow \theta$.

If θ is T, then φ is T, φ' is φ, θ' is θ. By inductive hypothesis, $\mathscr{B}'_1, \ldots, \mathscr{B}'_n \vdash \theta$. By $\prod_{A,B}^{\theta,\psi}$ (I.1) and DR, $\mathscr{B}'_1, \ldots, \mathscr{B}'_n \vdash \psi \Rightarrow \theta$. But φ' is $\psi \Rightarrow \theta$.

If ψ is T and θ is F, then φ is F, φ' is $\bar{\varphi}$, ψ' is ψ, and θ' is $\bar{\theta}$. By inductive hypothesis, $\mathscr{B}'_1 \ldots, \mathscr{B}'_n \vdash \psi$ and $\mathscr{B}'_1, \ldots, \mathscr{B}'_n \vdash \bar{\theta}$. Hence, by Lemma 1(f), $\mathscr{B}'_1, \ldots, \mathscr{B}'_n \vdash \overline{\psi \Rightarrow \theta}$. But φ' is $\overline{\psi \Rightarrow \theta}$.

COMPLETENESS THEOREM
Any tautology φ is a derivable formula.

PROOF
Let the variables of φ be $\mathscr{B}_1, \ldots, \mathscr{B}_n$. For any assignment of truth values, φ is T, and, therefore, φ' is φ. (We are using the notation of Lemma 2.) Hence, by Lemma 2, $\mathscr{B}'_1, \ldots, \mathscr{B}'_n \vdash \varphi$. If we choose $\mathscr{B}_n$ to be T, we have $\mathscr{B}'_1, \ldots, \mathscr{B}'_{n-1}, \mathscr{B}_n \vdash \varphi$, and, hence, by the deduction theorem (Appendix II),

$$\mathscr{B}'_1, \ldots, \mathscr{B}'_{n-1} \vdash \mathscr{B}_n \Rightarrow \varphi. \tag{*}$$

If we choose $\mathscr{B}_n$ to be F, we have $\mathscr{B}'_1, \ldots, \mathscr{B}'_{n-1}, \overline{\mathscr{B}_n} \vdash \varphi$, and, hence, by the deduction theorem,

$$\mathscr{B}'_1, \ldots, \mathscr{B}'_{n-1} \vdash \overline{\mathscr{B}_n} \Rightarrow \varphi. \tag{**}$$

By Lemma 1(h), (*), and (**), we obtain
$\mathscr{B}'_1, \ldots, \mathscr{B}'_{n-1} \vdash \varphi$.

Applying to $\mathscr{B}_{n-1}$ precisely the same procedure we just applied to $\mathscr{B}_n$, we can eliminate $\mathscr{B}_{n-1}$, and obtain $\mathscr{B}'_1, \ldots, \mathscr{B}'_{n-2} \vdash \varphi$, etc., until we finally obtain $\vdash \varphi$.

BIBLIOGRAPHY

1. Church, A. *Introduction to Mathematical Logic*, I, Princeton University Press, 1956.
2. Goodstein, R. L. *Mathematical Logic*, New York, Ungar, 1961.
3. Hilbert, D. and W. Ackermann, *Principles of Mathematical Logic*, Chelsea, New York, 1950 (translation of 1928 German edition).
4. Hohn, F. *Applied Boolean Algebra*, Macmillan, New York, 1960.
5. Kleene, S. C. *Introduction to Metamathematics*, Van Nostrand, Princeton, 1952.
6. Mendelson, E. *Introduction to Mathematical Logic*, Van Nostrand, Princeton, 1964.
7. Mendelson, E. *Boolean Algebra and Switching Circuits*, McGraw-Hill (Schaum), New York, 1970.
8. Novikov, P. S. *Elements of Mathematical Logic*, Addison-Wesley, 1964 (translation of 1959 Russian edition).
9. Stoll, R. R. *Sets, Logic, and Axiomatic Theories*, Freeman, San Francisco, 1961.
10. Tarski, A. *Introduction to Logic and to the Methodology of Deductive Sciences*, 3rd Edition, Oxford University Press, New York, 1965.

INDEX OF SPECIAL SYMBOLS

$\in$	13	dnf	79
$x \xrightarrow{f} y$	19	NOT	90
$x \to f(x)$	19	OR	90
$M \xrightarrow{f} N$	19	AND	90
$M \xrightarrow{f} \!\!\!\!\to N$	20	$\prod_A^\psi(\varphi)$	106
M^2	22	**DR**	106
M^3	23	$\prod_{A_1;A_2;\ldots;A_n}^{\varphi_1;\varphi_2;\ldots;\varphi_n}(\varphi)$	108
T	26	**CDR**	111
F	26	$\vdash$	112
$\overline{X}$	34	$\subset$	129
$\wedge$	35	$\subseteq$	130
$\bigwedge_{i=1}^{n}$	35	$\varnothing$	131
$\vee$	37	U	131
$\bigvee_{i=1}^{n}$	37	$\cap$	132
$\dot{\vee}$	38	$\cup$	132
$\Rightarrow$	39	$\forall$	149
$\Leftrightarrow$	42	$\exists$	149
$\vDash$	58, 163	$(\forall x)_{\in N}$	154
$\oplus$	73		

INDEX OF SPECIAL SYMBOLS

INDEX

Absolute geometry, 127
Ackermann, W., 6
Adder, single-bit binary, 90
Algebra
  Boolean, 73
  of logic, 3
  of propositions, 18
Algorithm, 61
Alphabet (of a calculus), 98
Analysis of discrete-action networks, 82
And, 34–35
AND, 90
Argument (of a function), 19
Aristotle, 2, 3
Arrangements with repetitions, 24
Associative, 53
Associativity
  of conjunction, 59
  of disjunction, 59
Atomic formulas, 103
Axiomatic method, 96
Axiom system for propositional calculus, 105

Barbara (syllogism mode), 174
Baroko (syllogism mode), 177
Beltrami, E., 122
Bernays, P., 6
Binary adder, single-bit, 90
Bolyai, J., 4
Boole, G., 3
Boolean
  algebra, 73
  functions, 73
Bound variables, 149, 156–157
Brouwer, L. E. J., 6

Calculus, 98
  axiom system for propositional, 105
  extended predicate, 152
  functional, 19
  predicate, 19, 166
  propositional, 18, 102
  restricted predicate, 152
Camestres (syllogism mode), 177

Categorical syllogism, 169
  modes of a, 170
CDR (composite detachment rule), 111
Celarent (syllogism mode), 175
Cesare (syllogism mode), 177
Church, A., 6, 167
Church's Theorem, 167
Combination of premises, rule for, 115
Commutative, 53
Commutativity
  of conjunction, 59
  of disjunction, 59
Complement, 131
Complete
  in the broad sense, 123
  in the narrow sense, 123
Completeness
  of an axiom system, 122–124
  of the propositional calculus, 123, 193
  theorem for the propositional calculus, 193
Composite detachment rule, 111
Compound proposition, 15, 30
Conclusion, 38
Conditional proposition, 38
Conjunction, 33–35
  associativity of, 59
  commutativity of, 59
  distributivity of conjunction with respect to disjunction, 59
  distributivity of disjunction with respect to conjunction, 60
  elementary, 76
Conjunctive
  normal form, 80
  perfect conjunctive normal form, 75
Consequence, 38
Consistency, 115
Constant, individual, 130
Contact network, 83
Contactless network, 89
Contensive, 72
Contingent, 58
Contradiction, law of, 60

204 INDEX

Darii (syllogism mode), 176
Decision
  method, 61
  problem, 61
  procedure, 61
Deduction Theorem, 113
  proof of, 191
de Morgan, A., 3
de Morgan's Laws, 60
Derivation (in proposition
  calculus), 107, 112
  rule, 63
Derived
  formula, 99
  proposition (in predicate logic), 182
Detachment rule, 106
  composite, 111
Discrete-action network, 82
Disjoint, 135
Disjunction, 36, 37
  associativity of, 59
  commutativity of, 59
  distributivity of conjunction
    with respect to disjunction, 59
  distributivity of disjunction with
    respect to conjunction, 60
  elementary, 76
  strict, 37
Disjunctive
  minimal disjunctive normal
    form, 80
  normal form, 79
  perfect disjunctive normal form, 75
Distributive, 53
dnf (disjunctive normal form), 79
Domain of definition of a
  function, 19
Double negation, law of, 59
DR (detachment rule), 106
Dual, 53
Duality law (principle), 53
  proof of, 190

Ehrenfest, P., 83
Element (of a set), 128

Elementary
  conjunction, 76
  disjunction, 76
  formula, 103
  proposition, 15, 30
Elements, functional, 90
Empty set, 131
Entailment, 41
Equality, 178
  predicate logic with, 180
  of sets, 129
equiv, 49, 158
Equivalence, 42–44
Equivalent formulas, 48, 158
Euclidean geometry, 5, 96–97, 120–122
Excluded middle, law of the, 60, 160
Existential quantifier, 149
Extended contraposition
  law, 61
  rule of, 68
Extended predicate calculus, 152

Factors, 73
False, identically, 58
Ferio (syllogism mode), 176
Festino (syllogism mode), 177
Formal logic, 2
Formalized language, 98, 99
Forms, propositional, 17
Formula
  of the algebra of propositions, 46
  atomic (elementary), 103
  of a calculus, 99
  derived, 99
  identically false, 58
  identically true, 58
  of predicate calculus, 166
  of predicate logic, 156
  structural formula of a network, 86
Free variable, 153, 156–157
Frege, G., 6
Function, 19
  Boolean, 73
  domain of definition of a, 19
  logical, 26
  propositional, 27

Function—cont.
  range of values of a, 19
Functional
  calculus, 19
  elements, 90

Gentzen, G., 6
Geometry, 4–5, 120–122
Gödel, K., 6

Heyting, A., 6
Hilbert, D., 6, 96, 117
Hypotheses, 112

Idempotency laws, 60
Identically
  false, 58
  true, 58
if and only if, 42
if..., then..., 38
Implication, 38–42
Improper subset, 130
Included, 129
  properly, 129
Independent (axiom, axiom system), 119
Individual
  constant, 130
  variable, 130
Inference, 63
Initial substitution theorem, 109–110
Instance, substitution, 109–110
Interpretable (axiom system), 116
Interpretation (of a theory), 97, 99
Intersection, 132
Into (mapping), 19

Jevons, W. S., 3

Kleene, S. C., 6
Klein, F., 122
Kolmogorov, A. N., 6

Language
  formalized, 98, 99
  object, 48
Law
  of associativity of conjunction, 59
  of associativity of disjunction, 59
  of commutativity of conjunction, 59
  of commutativity of disjunction, 59
  of contradiction, 60
  contraposition, 60
  of distributivity of conjunction with respect to disjunction, 59
  of distributivity of disjunction with respect to conjunction, 60
  of double negation, 59
  duality, 53, 190
  of the excluded middle, 60, 160
  extended contraposition, 61
  identity, 60
  syllogism, 61
Laws
  de Morgan's, 60
  idempotency, 60
  of logic, 58
Leibniz, G. W., 3
Lobachevski, N. I., 4
Logic
  algebra of, 3
  formal, 2
  formula of predicate, 156
  laws of, 58
  mathematical, 2, 6–7
  predicate, 18–19, 137
  predicate logic with equality, 180
  propositional, 18
  restricted predicate, 152
  traditional, 169
Logical
  analysis of discrete-action networks, 82
  function, 26
  product, 35
  synthesis of discrete-action networks, 82
Lukasiewicz, J., 6

Many-place predicate, 140
Mapping, 19
  into, 19
  onto, 19

Markov, A. A., 6
Mathematical logic, 2, 6–7
Meaningful, 72
Metalanguage, 48
Method, decision, 61
Middle, law of the excluded, 60, 160
Minimal disjunctive normal form, 80
Minimization problem, 80
Model (of a theory), 197
Modes of a categorical syllogism, 170
Modus ponens, 65, 106
Modus tollens, 65

Negation, 33–34
  law of double, 59
Network
  contact, 83
  contactless, 89
  discrete-action, 82
  P-network, 86
  realization, 86
Non-Euclidean geometry, 4–5, 120–122
Normal form
  conjunctive, 80
  disjunctive, 79
  minimal disjunctive, 80
  perfect conjunctive, 75
  perfect disjunctive, 75
Not, 33
NOT, 90
Novikov, P. S., 6, 105
Numerical variables, 14

Object language, 48
One-place predicate, 139
Onto (mapping), 19
Or, 36
OR, 90
Ordered pair, 21
Overlap, 135

Pair, ordered, 21
Parallel, connected in, 85

Parentheses, 10
  conventions for omitting, 104
Peano, G., 6
Perfect
  conjunctive normal form, 75
  disjunctive normal form, 75
Permutation of premises, rule for, 115
P-network, 86
Poincaré, H., 122
Ponens, modus, 65, 106
Poretskiy, P. S., 3
Post, E. L., 6
Predicate
  calculus, 19, 166
  extended predicate calculus, 152
  formula of predicate logic, 156
  logic, 18, 19, 137
  logic with equality, 180
  many-place, 140
  one-place, 139
  restricted predicate calculus, 152
  variable, 139
  zero-place, 147
Premise, 38, 112
Principle, duality, 53, 190
Problem, decision, 61
Procedure, decision, 61
Product, 73
  logical, 35
Proof
  in predicate logic, 182
  in propositional calculus, 107
  tree, 108
Proper subset, 129
Propositional
  axiom system for propositional calculus, 105
  calculus, 18, 102
  forms, 17
  function, 27
  logic, 18
  variables, 17, 102
Propositions, 14
  algebra of, 18
  compound, 15, 30
  conditional, 38
  elementary, 15, 30

Quantification, 152
Quantifier
  existential, 149
  restricted universal, 154
  universal, 149
Quine, W. V. O., 6

Range of values
  of a function, 19
  of a variable, 13, 130
Rank (of elementary conjunctions or disjunctions), 76
Realizable (axiomatic system), 116
Restricted
  predicate logic (calculus), 152
  universal quantifier, 154
Rosser, J. B., 6
Rule
  for combination of premises, 115
  composite detachment, 111
  of contraposition, 67
  of derivation, 63
  detachment, 106
  of extended contraposition, 68
  for permutation of premises, 115
  substitution, 106
  syllogism, 69, 114
Russell, B., 6

Schröder, E., 3
Semantics, 115
Series, connected in, 85
Set, 128
  empty, 131
  solution, 142
  truth, 139
  universal, 131
Shanin, N. S., 6
Shannon, C. E., 83
Shestakov, V. I., 83
Single-bit binary adder, 90
Skolem, T., 6
Solution set, 142
Statement, 8
Strict disjunction, 37
Structural formula of a network, 86

Subset, 129
  improper, 130
  proper, 129
Substitution
  initial substitution theorem, 109–110
  instance, 109–110
  rule, 106
Sum, 73
Summands, 73
Syllogism
  categorical, 169
  law, 61
  modes of a, 170
  rule, 69, 114
Syntax, 115
Synthesis, logical (of discrete-action networks), 82

Table, truth, 34
Tarski, A., 6
Tautology, 58
Terms (of a categorical syllogism), 169
Theorem
  deduction, 113, 191
  initial substitution, 109–110
  of predicate logic, 182
Tollens, modus, 65
Traditional logic, 169
Tree, proof, 108
True, identically, 58
Truth
  set, 139
  table, 34
  values, 32
Turing, A. M., 6

Union, 132
Universal
  quantifier, 149
  restricted universal quantifier, 154
  set, 131
Universally valid, 163

Valid, universally, 163

Variables, 12, 13
  bound, 149, 156–157
  free, 153, 156–157
  individual, 130
  numerical, 14
  predicate, 139
  propositional, 17, 102
  range of values of, 130
  values of, 130
Values
  range of values of variables, 13, 130
  truth, 32
  of variables, 13, 130

Whitehead, A. N., 6

Zero-place predicate, 147

A CATALOGUE OF SELECTED DOVER BOOKS IN ALL FIELDS OF INTEREST

A CATALOGUE OF SELECTED DOVER BOOKS IN ALL FIELDS OF INTEREST

RACKHAM'S COLOR ILLUSTRATIONS FOR WAGNER'S RING. Rackham's finest mature work—all 64 full-color watercolors in a faithful and lush interpretation of the *Ring*. Full-sized plates on coated stock of the paintings used by opera companies for authentic staging of Wagner. Captions aid in following complete Ring cycle. Introduction. 64 illustrations plus vignettes. 72pp. 8⅝ x 11¼. 23779-6 Pa. $6.00

CONTEMPORARY POLISH POSTERS IN FULL COLOR, edited by Joseph Czestochowski. 46 full-color examples of brilliant school of Polish graphic design, selected from world's first museum (near Warsaw) dedicated to poster art. Posters on circuses, films, plays, concerts all show cosmopolitan influences, free imagination. Introduction. 48pp. 9⅜ x 12¼. 23780-X Pa. $6.00

GRAPHIC WORKS OF EDVARD MUNCH, Edvard Munch. 90 haunting, evocative prints by first major Expressionist artist and one of the greatest graphic artists of his time: *The Scream, Anxiety, Death Chamber, The Kiss, Madonna*, etc. Introduction by Alfred Werner. 90pp. 9 x 12. 23765-6 Pa. $5.00

THE GOLDEN AGE OF THE POSTER, Hayward and Blanche Cirker. 70 extraordinary posters in full colors, from Maitres de l'Affiche, Mucha, Lautrec, Bradley, Cheret, Beardsley, many others. Total of 78pp. 9⅜ x 12¼. 22753-7 Pa. $6.95

THE NOTEBOOKS OF LEONARDO DA VINCI, edited by J. P. Richter. Extracts from manuscripts reveal great genius; on painting, sculpture, anatomy, sciences, geography, etc. Both Italian and English. 186 ms. pages reproduced, plus 500 additional drawings, including studies for *Last Supper*, Sforza monument, etc. 860pp. 7⅞ x 10¾. (Available in U.S. only) 22572-0, 22573-9 Pa., Two-vol. set $19.90

THE CODEX NUTTALL, as first edited by Zelia Nuttall. Only inexpensive edition, in full color, of a pre-Columbian Mexican (Mixtec) book. 88 color plates show kings, gods, heroes, temples, sacrifices. New explanatory, historical introduction by Arthur G. Miller. 96pp. 11⅜ x 8½. (Available in U.S. only) 23168-2 Pa. $7.95

UNE SEMAINE DE BONTÉ, A SURREALISTIC NOVEL IN COLLAGE, Max Ernst. Masterpiece created out of 19th-century periodical illustrations, explores worlds of terror and surprise. Some consider this Ernst's greatest work. 208pp. 8⅛ x 11. 23252-2 Pa. $6.00

CATALOGUE OF DOVER BOOKS

DRAWINGS OF WILLIAM BLAKE, William Blake. 92 plates from Book of Job, *Divine Comedy, Paradise Lost*, visionary heads, mythological figures, Laocoon, etc. Selection, introduction, commentary by Sir Geoffrey Keynes. 178pp. 8⅛ x 11. 22303-5 Pa. $5.00

ENGRAVINGS OF HOGARTH, William Hogarth. 101 of Hogarth's greatest works: *Rake's Progress, Harlot's Progress, Illustrations for Hudibras, Before and After, Beer Street and Gin Lane*, many more. Full commentary. 256pp. 11 x 13¾. 22479-1 Pa. $12.95

DAUMIER: 120 GREAT LITHOGRAPHS, Honore Daumier. Wide-ranging collection of lithographs by the greatest caricaturist of the 19th century. Concentrates on eternally popular series on lawyers, on married life, on liberated women, etc. Selection, introduction, and notes on plates by Charles F. Ramus. Total of 158pp. 9⅜ x 12¼. 23512-2 Pa. $6.00

DRAWINGS OF MUCHA, Alphonse Maria Mucha. Work reveals draftsman of highest caliber: studies for famous posters and paintings, renderings for book illustrations and ads, etc. 70 works, 9 in color; including 6 items not drawings. Introduction. List of illustrations. 72pp. 9⅜ x 12¼. (Available in U.S. only) 23672-2 Pa. $4.50

GIOVANNI BATTISTA PIRANESI: DRAWINGS IN THE PIERPONT MORGAN LIBRARY, Giovanni Battista Piranesi. For first time ever all of Morgan Library's collection, world's largest. 167 illustrations of rare Piranesi drawings—archeological, architectural, decorative and visionary. Essay, detailed list of drawings, chronology, captions. Edited by Felice Stampfle. 144pp. 9⅜ x 12¼. 23714-1 Pa. $7.50

NEW YORK ETCHINGS (1905-1949), John Sloan. All of important American artist's N.Y. life etchings. 67 works include some of his best art; also lively historical record—Greenwich Village, tenement scenes. Edited by Sloan's widow. Introduction and captions. 79pp. 8⅜ x 11¼.
23651-X Pa. $5.00

CHINESE PAINTING AND CALLIGRAPHY: A PICTORIAL SURVEY, Wan-go Weng. 69 fine examples from John M. Crawford's matchless private collection: landscapes, birds, flowers, human figures, etc., plus calligraphy. Every basic form included: hanging scrolls, handscrolls, album leaves, fans, etc. 109 illustrations. Introduction. Captions. 192pp. 8⅞ x 11¾.
23707-9 Pa. $7.95

DRAWINGS OF REMBRANDT, edited by Seymour Slive. Updated Lippmann, Hofstede de Groot edition, with definitive scholarly apparatus. All portraits, biblical sketches, landscapes, nudes, Oriental figures, classical studies, together with selection of work by followers. 550 illustrations. Total of 630pp. 9⅛ x 12¼. 21485-0, 21486-9 Pa., Two-vol. set $17.90

THE DISASTERS OF WAR, Francisco Goya. 83 etchings record horrors of Napoleonic wars in Spain and war in general. Reprint of 1st edition, plus 3 additional plates. Introduction by Philip Hofer. 97pp. 9⅜ x 8¼.
21872-4 Pa. $4.50

CATALOGUE OF DOVER BOOKS

THE EARLY WORK OF AUBREY BEARDSLEY, Aubrey Beardsley. 157 plates, 2 in color: *Manon Lescaut, Madame Bovary, Morte Darthur, Salome,* other. Introduction by H. Marillier. 182pp. 8⅛ x 11. 21816-3 Pa. $6.50

THE LATER WORK OF AUBREY BEARDSLEY, Aubrey Beardsley. Exotic masterpieces of full maturity: *Venus and Tannhauser, Lysistrata, Rape of the Lock, Volpone,* Savoy material, etc. 174 plates, 2 in color. 186pp. 8⅛ x 11. 21817-1 Pa. $5.95

THOMAS NAST'S CHRISTMAS DRAWINGS, Thomas Nast. Almost all Christmas drawings by creator of image of Santa Claus as we know it, and one of America's foremost illustrators and political cartoonists. 66 illustrations. 3 illustrations in color on covers. 96pp. 8⅜ x 11¼.
23660-9 Pa. $3.50

THE DORÉ ILLUSTRATIONS FOR DANTE'S DIVINE COMEDY, Gustave Doré. All 135 plates from Inferno, Purgatory, Paradise; fantastic tortures, infernal landscapes, celestial wonders. Each plate with appropriate (translated) verses. 141pp. 9 x 12. 23231-X Pa. $5.00

DORÉ'S ILLUSTRATIONS FOR RABELAIS, Gustave Doré. 252 striking illustrations of *Gargantua and Pantagruel* books by foremost 19th-century illustrator. Including 60 plates, 192 delightful smaller illustrations. 153pp. 9 x 12. 23656-0 Pa. $6.00

LONDON: A PILGRIMAGE, Gustave Doré, Blanchard Jerrold. Squalor, riches, misery, beauty of mid-Victorian metropolis; 55 wonderful plates, 125 other illustrations, full social, cultural text by Jerrold. 191pp. of text. 9⅜ x 12¼. 22306-X Pa. $7.00

THE RIME OF THE ANCIENT MARINER, Gustave Doré, S. T. Coleridge. Dore's finest work, 34 plates capture moods, subtleties of poem. Full text. Introduction by Millicent Rose. 77pp. 9¼ x 12. 22305-1 Pa. $4.50

THE DORE BIBLE ILLUSTRATIONS, Gustave Doré. All wonderful, detailed plates: Adam and Eve, Flood, Babylon, Life of Jesus, etc. Brief King James text with each plate. Introduction by Millicent Rose. 241 plates. 241pp. 9 x 12. 23004-X Pa. $6.95

THE COMPLETE ENGRAVINGS, ETCHINGS AND DRYPOINTS OF ALBRECHT DURER. "Knight, Death and Devil"; "Melencolia," and more—all Dürer's known works in all three media, including 6 works formerly attributed to him. 120 plates. 235pp. 8⅜ x 11¼.
22851-7 Pa. $7.50

MECHANICK EXERCISES ON THE WHOLE ART OF PRINTING, Joseph Moxon. First complete book (1683-4) ever written about typography, a compendium of everything known about printing at the latter part of 17th century. Reprint of 2nd (1962) Oxford Univ. Press edition. 74 illustrations. Total of 550pp. 6⅛ x 9¼. 23617-X Pa. $7.95

CATALOGUE OF DOVER BOOKS

THE COMPLETE WOODCUTS OF ALBRECHT DURER, edited by Dr. W. Kurth. 346 in all: "Old Testament," "St. Jerome," "Passion," "Life of Virgin," Apocalypse," many others. Introduction by Campbell Dodgson. 285pp. 8½ x 12¼. 21097-9 Pa. $7.50

DRAWINGS OF ALBRECHT DURER, edited by Heinrich Wolfflin. 81 plates show development from youth to full style. Many favorites; many new. Introduction by Alfred Werner. 96pp. 8⅛ x 11. 22352-3 Pa. $6.00

THE HUMAN FIGURE, Albrecht Dürer. Experiments in various techniques—stereometric, progressive proportional, and others. Also life studies that rank among finest ever done. Complete reprinting of *Dresden Sketchbook*. 170 plates. 355pp. 8⅜ x 11¼. 21042-1 Pa. $7.95

OF THE JUST SHAPING OF LETTERS, Albrecht Dürer. Renaissance artist explains design of Roman majuscules by geometry, also Gothic lower and capitals. Grolier Club edition. 43pp. 7⅞ x 10¾ 21306-4 Pa. $3.00

TEN BOOKS ON ARCHITECTURE, Vitruvius. The most important book ever written on architecture. Early Roman aesthetics, technology, classical orders, site selection, all other aspects. Stands behind everything since. Morgan translation. 331pp. 5⅜ x 8½. 20645-9 Pa. $5.00

THE FOUR BOOKS OF ARCHITECTURE, Andrea Palladio. 16th-century classic responsible for Palladian movement and style. Covers classical architectural remains, Renaissance revivals, classical orders, etc. 1738 Ware English edition. Introduction by A. Placzek. 216 plates. 110pp. of text. 9½ x 12¾. 21308-0 Pa. $10.00

HORIZONS, Norman Bel Geddes. Great industrialist stage designer, "father of streamlining," on application of aesthetics to transportation, amusement, architecture, etc. 1932 prophetic account; function, theory, specific projects. 222 illustrations. 312pp. 7⅞ x 10¾. 23514-9 Pa. $6.95

FRANK LLOYD WRIGHT'S FALLINGWATER, Donald Hoffmann. Full, illustrated story of conception and building of Wright's masterwork at Bear Run, Pa. 100 photographs of site, construction, and details of completed structure. 112pp. 9¼ x 10. 23671-4 Pa. $5.95

THE ELEMENTS OF DRAWING, John Ruskin. Timeless classic by great Viltorian; starts with basic ideas, works through more difficult. Many practical exercises. 48 illustrations. Introduction by Lawrence Campbell. 228pp. 5⅜ x 8½. 22730-8 Pa. $3.75

GIST OF ART, John Sloan. Greatest modern American teacher, Art Students League, offers innumerable hints, instructions, guided comments to help you in painting. Not a formal course. 46 illustrations. Introduction by Helen Sloan. 200pp. 5⅜ x 8½. 23435-5 Pa. $4.00

CATALOGUE OF DOVER BOOKS

THE ANATOMY OF THE HORSE, George Stubbs. Often considered the great masterpiece of animal anatomy. Full reproduction of 1766 edition, plus prospectus; original text and modernized text. 36 plates. Introduction by Eleanor Garvey. 121pp. 11 x 14¾. 23402-9 Pa. $8.95

BRIDGMAN'S LIFE DRAWING, George B. Bridgman. More than 500 illustrative drawings and text teach you to abstract the body into its major masses, use light and shade, proportion; as well as specific areas of anatomy, of which Bridgman is master. 192pp. 6½ x 9¼. (Available in U.S. only) 22710-3 Pa. $4.50

ART NOUVEAU DESIGNS IN COLOR, Alphonse Mucha, Maurice Verneuil, Georges Auriol. Full-color reproduction of *Combinaisons ornementales* (c. 1900) by Art Nouveau masters. Floral, animal, geometric, interlacings, swashes—borders, frames, spots—all incredibly beautiful. 60 plates, hundreds of designs. 9⅜ x 8-1/16. 22885-1 Pa. $4.50

FULL-COLOR FLORAL DESIGNS IN THE ART NOUVEAU STYLE, E. A. Seguy. 166 motifs, on 40 plates, from *Les fleurs et leurs applications decoratives* (1902): borders, circular designs, repeats, allovers, "spots." All in authentic Art Nouveau colors. 48pp. 9⅜ x 12¼. 23439-8 Pa. $5.00

A DIDEROT PICTORIAL ENCYCLOPEDIA OF TRADES AND INDUSTRY, edited by Charles C. Gillispie. 485 most interesting plates from the great French Encyclopedia of the 18th century show hundreds of working figures, artifacts, process, land and cityscapes; glassmaking, papermaking, metal extraction, construction, weaving, making furniture, clothing, wigs, dozens of other activities. Plates fully explained. 920pp. 9 x 12. 22284-5, 22285-3 Clothbd., Two-vol. set $40.00

HANDBOOK OF EARLY ADVERTISING ART, Clarence P. Hornung. Largest collection of copyright-free early and antique advertising art ever compiled. Over 6,000 illustrations, from Franklin's time to the 1890's for special effects, novelty. Valuable source, almost inexhaustible.
Pictorial Volume. Agriculture, the zodiac, animals, autos, birds, Christmas, fire engines, flowers, trees, musical instruments, ships, games and sports, much more. Arranged by subject matter and use. 237 plates. 288pp. 9 x 12. 20122-8 Clothbd. $15.00

Typographical Volume. Roman and Gothic faces ranging from 10 point to 300 point, "Barnum," German and Old English faces, script, logotypes, scrolls and flourishes, 1115 ornamental initials, 67 complete alphabets, more. 310 plates. 320pp. 9 x 12. 20123-6 Clothbd. $15.00

CALLIGRAPHY (CALLIGRAPHIA LATINA), J. G. Schwandner. High point of 18th-century ornamental calligraphy. Very ornate initials, scrolls, borders, cherubs, birds, lettered examples. 172pp. 9 x 13. 20475-8 Pa. $7.95

CATALOGUE OF DOVER BOOKS

ART FORMS IN NATURE, Ernst Haeckel. Multitude of strangely beautiful natural forms: Radiolaria, Foraminifera, jellyfishes, fungi, turtles, bats, etc. All 100 plates of the 19th-century evolutionist's *Kunstformen der Natur* (1904). 100pp. 9⅜ x 12¼. 22987-4 Pa. $5.00

CHILDREN: A PICTORIAL ARCHIVE FROM NINETEENTH-CENTURY SOURCES, edited by Carol Belanger Grafton. 242 rare, copyright-free wood engravings for artists and designers. Widest such selection available. All illustrations in line. 119pp. 8⅜ x 11¼.
23694-3 Pa. $4.00

WOMEN: A PICTORIAL ARCHIVE FROM NINETEENTH-CENTURY SOURCES, edited by Jim Harter. 391 copyright-free wood engravings for artists and designers selected from rare periodicals. Most extensive such collection available. All illustrations in line. 128pp. 9 x 12.
23703-6 Pa. $4.95

ARABIC ART IN COLOR, Prisse d'Avennes. From the greatest ornamentalists of all time—50 plates in color, rarely seen outside the Near East, rich in suggestion and stimulus. Includes 4 plates on covers. 46pp. 9⅜ x 12¼. 23658-7 Pa. $6.00

AUTHENTIC ALGERIAN CARPET DESIGNS AND MOTIFS, edited by June Beveridge. Algerian carpets are world famous. Dozens of geometrical motifs are charted on grids, color-coded, for weavers, needleworkers, craftsmen, designers. 53 illustrations plus 4 in color. 48pp. 8¼ x 11. (Available in U.S. only) 23650-1 Pa. $1.75

DICTIONARY OF AMERICAN PORTRAITS, edited by Hayward and Blanche Cirker. 4000 important Americans, earliest times to 1905, mostly in clear line. Politicians, writers, soldiers, scientists, inventors, industrialists, Indians, Blacks, women, outlaws, etc. Identificatory information. 756pp. 9¼ x 12¾. 21823-6 Clothbd. $65.00

HOW THE OTHER HALF LIVES, Jacob A. Riis. Journalistic record of filth, degradation, upward drive in New York immigrant slums, shops, around 1900. New edition includes 100 original Riis photos, monuments of early photography. 233pp. 10 x 7⅞. 22012-5 Pa. $7.00

NEW YORK IN THE THIRTIES, Berenice Abbott. Noted photographer's fascinating study of city shows new buildings that have become famous and old sights that have disappeared forever. Insightful commentary. 97 photographs. 97pp. 11⅜ x 10. 22967-X Pa. $6.00

MEN AT WORK, Lewis W. Hine. Famous photographic studies of construction workers, railroad men, factory workers and coal miners. New supplement of 18 photos on Empire State building construction. New introduction by Jonathan L. Doherty. Total of 69 photos. 63pp. 8 x 10¾.
23475-4 Pa. $4.00

CATALOGUE OF DOVER BOOKS

THE DEPRESSION YEARS AS PHOTOGRAPHED BY ARTHUR ROTHSTEIN, Arthur Rothstein. First collection devoted entirely to the work of outstanding 1930s photographer: famous dust storm photo, ragged children, unemployed, etc. 120 photographs. Captions. 119pp. 9¼ x 10¾.
23590-4 Pa. **$5.95**

CAMERA WORK: A PICTORIAL GUIDE, Alfred Stieglitz. All 559 illustrations and plates from the most important periodical in the history of art photography, Camera Work (1903-17). Presented four to a page, reduced in size but still clear, in strict chronological order, with complete captions. Three indexes. Glossary. Bibliography. 176pp. 8⅜ x 11¼.
23591-2 Pa. **$6.95**

ALVIN LANGDON COBURN, PHOTOGRAPHER, Alvin L. Coburn. Revealing autobiography by one of greatest photographers of 20th century gives insider's version of Photo-Secession, plus comments on his own work. 77 photographs by Coburn. Edited by Helmut and Alison Gernsheim. 160pp. 8⅛ x 11.
23685-4 Pa. **$6.00**

NEW YORK IN THE FORTIES, Andreas Feininger. 162 brilliant photographs by the well-known photographer, formerly with Life magazine, show commuters, shoppers, Times Square at night, Harlem nightclub, Lower East Side, etc. Introduction and full captions by John von Hartz. 181pp. 9¼ x 10¾.
23585-8 Pa. **$6.95**

GREAT NEWS PHOTOS AND THE STORIES BEHIND THEM, John Faber. Dramatic volume of 140 great news photos, 1855 through 1976, and revealing stories behind them, with both historical and technical information. Hindenburg disaster, shooting of Oswald, nomination of Jimmy Carter, etc. 160pp. 8¼ x 11.
23667-6 Pa. **$6.00**

THE ART OF THE CINEMATOGRAPHER, Leonard Maltin. Survey of American cinematography history and anecdotal interviews with 5 masters—Arthur Miller, Hal Mohr, Hal Rosson, Lucien Ballard, and Conrad Hall. Very large selection of behind-the-scenes production photos. 105 photographs. Filmographies. Index. Originally Behind the Camera. 144pp. 8¼ x 11.
23686-2 Pa. **$5.00**

DESIGNS FOR THE THREE-CORNERED HAT (LE TRICORNE), Pablo Picasso. 32 fabulously rare drawings—including 31 color illustrations of costumes and accessories—for 1919 production of famous ballet. Edited by Parmenia Migel, who has written new introduction. 48pp. 9⅜ x 12¼. (Available in U.S. only)
23709-5 Pa. **$5.00**

NOTES OF A FILM DIRECTOR, Sergei Eisenstein. Greatest Russian filmmaker explains montage, making of Alexander Nevsky, aesthetics; comments on self, associates, great rivals (Chaplin), similar material. 78 illustrations. 240pp. 5⅜ x 8½.
22392-2 Pa. **$7.00**

CATALOGUE OF DOVER BOOKS

HOLLYWOOD GLAMOUR PORTRAITS, edited by John Kobal. 145 photos capture the stars from 1926-49, the high point in portrait photography. Gable, Harlow, Bogart, Bacall, Hedy Lamarr, Marlene Dietrich, Robert Montgomery, Marlon Brando, Veronica Lake; 94 stars in all. Full background on photographers, technical aspects, much more. Total of 160pp. 8⅜ x 11¼. 23352-9 Pa. $6.95

THE NEW YORK STAGE: FAMOUS PRODUCTIONS IN PHOTOGRAPHS, edited by Stanley Appelbaum. 148 photographs from Museum of City of New York show 142 plays, 1883-1939. *Peter Pan, The Front Page, Dead End, Our Town,* O'Neill, hundreds of actors and actresses, etc. Full indexes. 154pp. 9½ x 10. 23241-7 Pa. $6.00

DIALOGUES CONCERNING TWO NEW SCIENCES, Galileo Galilei. Encompassing 30 years of experiment and thought, these dialogues deal with geometric demonstrations of fracture of solid bodies, cohesion, leverage, speed of light and sound, pendulums, falling bodies, accelerated motion, etc. 300pp. 5⅜ x 8½. 60099-8 Pa. $5.50

THE GREAT OPERA STARS IN HISTORIC PHOTOGRAPHS, edited by James Camner. 343 portraits from the 1850s to the 1940s: Tamburini, Mario, Caliapin, Jeritza, Melchior, Melba, Patti, Pinza, Schipa, Caruso, Farrar, Steber, Gobbi, and many more—270 performers in all. Index. 199pp. 8⅜ x 11¼. 23575-0 Pa. $7.50

J. S. BACH, Albert Schweitzer. Great full-length study of Bach, life, background to music, music, by foremost modern scholar. Ernest Newman translation. 650 musical examples. Total of 928pp. 5⅜ x 8½. (Available in U.S. only) 21631-4, 21632-2 Pa., Two-vol. set $12.00

COMPLETE PIANO SONATAS, Ludwig van Beethoven. All sonatas in the fine Schenker edition, with fingering, analytical material. One of best modern editions. Total of 615pp. 9 x 12. (Available in U.S. only) 23134-8, 23135-6 Pa., Two-vol. set $17.90

KEYBOARD MUSIC, J. S. Bach. Bach-Gesellschaft edition. For harpsichord, piano, other keyboard instruments. English Suites, French Suites, Six Partitas, Goldberg Variations, Two-Part Inventions, Three-Part Sinfonias. 312pp. 8⅛ x 11. (Available in U.S. only) 22360-4 Pa. $7.95

FOUR SYMPHONIES IN FULL SCORE, Franz Schubert. Schubert's four most popular symphonies: No. 4 in C Minor ("Tragic"); No. 5 in B-flat Major; No. 8 in B Minor ("Unfinished"); No. 9 in C Major ("Great"). Breitkopf & Hartel edition. Study score. 261pp. 9⅜ x 12¼. 23681-1 Pa. $8.95

THE AUTHENTIC GILBERT & SULLIVAN SONGBOOK, W. S. Gilbert, A. S. Sullivan. Largest selection available; 92 songs, uncut, original keys, in piano rendering approved by Sullivan. Favorites and lesser-known fine numbers. Edited with plot synopses by James Spero. 3 illustrations. 399pp. 9 x 12. 23482-7 Pa. $10.95

CATALOGUE OF DOVER BOOKS

PRINCIPLES OF ORCHESTRATION, Nikolay Rimsky-Korsakov. Great classical orchestrator provides fundamentals of tonal resonance, progression of parts, voice and orchestra, tutti effects, much else in major document. 330pp. of musical excerpts. 489pp. 6½ x 9¼. 21266-1 Pa. $7.50

TRISTAN UND ISOLDE, Richard Wagner. Full orchestral score with complete instrumentation. Do not confuse with piano reduction. Commentary by Felix Mottl, great Wagnerian conductor and scholar. Study score. 655pp. 8⅛ x 11. 22915-7 Pa. $13.95

REQUIEM IN FULL SCORE, Giuseppe Verdi. Immensely popular with choral groups and music lovers. Republication of edition published by C. F. Peters, Leipzig, n. d. German frontmaker in English translation. Glossary. Text in Latin. Study score. 204pp. 9⅜ x 12¼.
23682-X Pa. $6.50

COMPLETE CHAMBER MUSIC FOR STRINGS, Felix Mendelssohn. All of Mendelssohn's chamber music: Octet, 2 Quintets, 6 Quartets, and Four Pieces for String Quartet. (Nothing with piano is included). Complete works edition (1874-7). Study score. 283 pp. 9⅜ x 12¼.
23679-X Pa. $7.50

POPULAR SONGS OF NINETEENTH-CENTURY AMERICA, edited by Richard Jackson. 64 most important songs: "Old Oaken Bucket," "Arkansas Traveler," "Yellow Rose of Texas," etc. Authentic original sheet music, full introduction and commentaries. 290pp. 9 x 12. 23270-0 Pa. $7.95

COLLECTED PIANO WORKS, Scott Joplin. Edited by Vera Brodsky Lawrence. Practically all of Joplin's piano works—rags, two-steps, marches, waltzes, etc., 51 works in all. Extensive introduction by Rudi Blesh. Total of 345pp. 9 x 12. 23106-2 Pa. $15.95

BASIC PRINCIPLES OF CLASSICAL BALLET, Agrippina Vaganova. Great Russian theoretician, teacher explains methods for teaching classical ballet; incorporates best from French, Italian, Russian schools. 118 illustrations. 175pp. 5⅜ x 8½. 22036-2 Pa. $2.75

CHINESE CHARACTERS, L. Wieger. Rich analysis of 2300 characters according to traditional systems into primitives. Historical-semantic analysis to phonetics (Classical Mandarin) and radicals. 820pp. 6⅛ x 9¼.
21321-8 Pa. $12.50

THE WARES OF THE MING DYNASTY, R. L. Hobson. Foremost scholar examines and illustrates many varieties of Ming (1368-1644). Famous blue and white, polychrome, lesser-known styles and shapes. 117 illustrations, 9 full color, of outstanding pieces. Total of 263pp. 6⅛ x 9¼. (Available in U.S. only) 23652-8 Pa. $6.00

AN ETYMOLOGICAL DICTIONARY OF MODERN ENGLISH, Ernest Weekley. Richest, fullest work, by foremost British lexicographer. Detailed word histories. Inexhaustible. Do not confuse this with *Concise Etymological Dictionary*, which is abridged. Total of 856pp. 6½ x 9¼.
21873-2, 21874-0 Pa., Two-vol. set $13.00

CATALOGUE OF DOVER BOOKS

A MAYA GRAMMAR, Alfred M. Tozzer. Practical, useful English-language grammar by the Harvard anthropologist who was one of the three greatest American scholars in the area of Maya culture. Phonetics, grammatical processes, syntax, more. 301pp. 5⅜ x 8½. 23465-7 Pa. $4.00

THE JOURNAL OF HENRY D. THOREAU, edited by Bradford Torrey, F. H. Allen. Complete reprinting of 14 volumes, 1837-61, over two million words; the sourcebooks for *Walden*, etc. Definitive. All original sketches, plus 75 photographs. Introduction by Walter Harding. Total of 1804pp. 8½ x 12¼. 20312-3, 20313-1 Clothbd., Two-vol. set $80.00

CLASSIC GHOST STORIES, Charles Dickens and others. 18 wonderful stories you've wanted to reread: "The Monkey's Paw," "The House and the Brain," "The Upper Berth," "The Signalman," "Dracula's Guest," "The Tapestried Chamber," etc. Dickens, Scott, Mary Shelley, Stoker, etc. 330pp. 5⅜ x 8½. 20735-8 Pa. $4.50

SEVEN SCIENCE FICTION NOVELS, H. G. Wells. Full novels. *First Men in the Moon, Island of Dr. Moreau, War of the Worlds, Food of the Gods, Invisible Man, Time Machine, In the Days of the Comet.* A basic science-fiction library. 1015pp. 5⅜ x 8½. (Available in U.S. only)
20264-X Clothbd. $15.00

ARMADALE, Wilkie Collins. Third great mystery novel by the author of *The Woman in White* and *The Moonstone*. Ingeniously plotted narrative shows an exceptional command of character, incident and mood. Original magazine version with 40 illustrations. 597pp. 5⅜ x 8½.
23429-0 Pa. $7.95

FLATLAND, E. A. Abbott. Science-fiction classic explores life of 2-D being in 3-D world. Read also as introduction to thought about hyperspace. Introduction by Banesh Hoffmann. 16 illustrations. 103pp. 5⅜ x 8½.
20001-9 Pa. $2.75

AYESHA: THE RETURN OF "SHE," H. Rider Haggard. Virtuoso sequel featuring the great mythic creation, Ayesha, in an adventure that is fully as good as the first book, *She*. Original magazine version, with 47 original illustrations by Maurice Greiffenhagen. 189pp. 6½ x 9¼.
23649-8 Pa. $3.50

ORIENTAL RUGS, ANTIQUE AND MODERN, Walter A. Hawley. Persia, Turkey, Caucasus, Central Asia, China, other traditions. Best general survey of all aspects: styles and periods, manufacture, uses, symbols and their interpretation, and identification. 96 illustrations, 11 in color. 320pp. 6⅛ x 9¼. 22366-3 Pa. $6.95

CHINESE POTTERY AND PORCELAIN, R. L. Hobson. Detailed descriptions and analyses by former Keeper of the Department of Oriental Antiquities and Ethnography at the British Museum. Covers hundreds of pieces from primitive times to 1915. Still the standard text for most periods. 136 plates, 40 in full color. Total of 750pp. 5⅜ x 8½.
23253-0 Pa. $10.00

CATALOGUE OF DOVER BOOKS

UNCLE SILAS, J. Sheridan LeFanu. Victorian Gothic mystery novel, considered by many best of period, even better than Collins or Dickens. Wonderful psychological terror. Introduction by Frederick Shroyer. 436pp. 5⅜ x 8½. 21715-9 Pa. $6.95

JURGEN, James Branch Cabell. The great erotic fantasy of the 1920's that delighted thousands, shocked thousands more. Full final text, Lane edition with 13 plates by Frank Pape. 346pp. 5⅜ x 8½.
23507-6 Pa. $4.50

THE CLAVERINGS, Anthony Trollope. Major novel, chronicling aspects of British Victorian society, personalities. Reprint of Cornhill serialization, 16 plates by M. Edwards; first reprint of full text. Introduction by Norman Donaldson. 412pp. 5⅜ x 8½. 23464-9 Pa. $5.00

KEPT IN THE DARK, Anthony Trollope. Unusual short novel about Victorian morality and abnormal psychology by the great English author. Probably the first American publication. Frontispiece by Sir John Millais. 92pp. 6½ x 9¼. 23609-9 Pa. $2.50

RALPH THE HEIR, Anthony Trollope. Forgotten tale of illegitimacy, inheritance. Master novel of Trollope's later years. Victorian country estates, clubs, Parliament, fox hunting, world of fully realized characters. Reprint of 1871 edition. 12 illustrations by F. A. Faser. 434pp. of text. 5⅜ x 8½. 23642-0 Pa. $6.50

YEKL and THE IMPORTED BRIDEGROOM AND OTHER STORIES OF THE NEW YORK GHETTO, Abraham Cahan. Film *Hester Street* based on *Yekl* (1896). Novel, other stories among first about Jewish immigrants of N.Y.'s East Side. Highly praised by W. D. Howells—Cahan "a new star of realism." New introduction by Bernard G. Richards. 240pp. 5⅜ x 8½. 22427-9 Pa. $3.50

THE HIGH PLACE, James Branch Cabell. Great fantasy writer's enchanting comedy of disenchantment set in 18th-century France. Considered by some critics to be even better than his famous *Jurgen*. 10 illustrations and numerous vignettes by noted fantasy artist Frank C. Pape. 320pp. 5⅜ x 8½. 23670-6 Pa. $4.00

ALICE'S ADVENTURES UNDER GROUND, Lewis Carroll. Facsimile of ms. Carroll gave Alice Liddell in 1864. Different in many ways from final Alice. Handlettered, illustrated by Carroll. Introduction by Martin Gardner. 128pp. 5⅜ x 8½. 21482-6 Pa. $2.50

FAVORITE ANDREW LANG FAIRY TALE BOOKS IN MANY COLORS, Andrew Lang. The four Lang favorites in a boxed set—the complete *Red, Green, Yellow* and *Blue* Fairy Books. 164 stories; 439 illustrations by Lancelot Speed, Henry Ford and G. P. Jacomb Hood. Total of about 1500pp. 5⅜ x 8½. 23407-X Boxed set, Pa. $16.95

CATALOGUE OF DOVER BOOKS

HOUSEHOLD STORIES BY THE BROTHERS GRIMM. All the great Grimm stories: "Rumpelstiltskin," "Snow White," "Hansel and Gretel," etc., with 114 illustrations by Walter Crane. 269pp. 5⅜ x 8½.
21080-4 Pa. $3.50

SLEEPING BEAUTY, illustrated by Arthur Rackham. Perhaps the fullest, most delightful version ever, told by C. S. Evans. Rackham's best work. 49 illustrations. 110pp. 7⅞ x 10¾.
22756-1 Pa. $2.95

AMERICAN FAIRY TALES, L. Frank Baum. Young cowboy lassoes Father Time; dummy in Mr. Floman's department store window comes to life; and 10 other fairy tales. 41 illustrations by N. P. Hall, Harry Kennedy, Ike Morgan, and Ralph Gardner. 209pp. 5⅜ x 8½.
23643-9 Pa. $3.00

THE WONDERFUL WIZARD OF OZ, L. Frank Baum. Facsimile in full color of America's finest children's classic. Introduction by Martin Gardner. 143 illustrations by W. W. Denslow. 267pp. 5⅜ x 8½.
20691-2 Pa. $4.50

THE TALE OF PETER RABBIT, Beatrix Potter. The inimitable Peter's terrifying adventure in Mr. McGregor's garden, with all 27 wonderful, full-color Potter illustrations. 55pp. 4¼ x 5½. (Available in U.S. only)
22827-4 Pa. $1.50

THE STORY OF KING ARTHUR AND HIS KNIGHTS, Howard Pyle. Finest children's version of life of King Arthur. 48 illustrations by Pyle. 131pp. 6⅛ x 9¼.
21445-1 Pa. $5.95

CARUSO'S CARICATURES, Enrico Caruso. Great tenor's remarkable caricatures of self, fellow musicians, composers, others. Toscanini, Puccini, Farrar, etc. Impish, cutting, insightful. 473 illustrations. Preface by M. Sisca. 217pp. 8⅜ x 11¼.
23528-9 Pa. $6.95

PERSONAL NARRATIVE OF A PILGRIMAGE TO ALMADINAH AND MECCAH, Richard Burton. Great travel classic by remarkably colorful personality. Burton, disguised as a Moroccan, visited sacred shrines of Islam, narrowly escaping death. Wonderful observations of Islamic life, customs, personalities. 47 illustrations. Total of 959pp. 5⅜ x 8½.
21217-3, 21218-1 Pa., Two-vol. set $14.00

INCIDENTS OF TRAVEL IN YUCATAN, John L. Stephens. Classic (1843) exploration of jungles of Yucatan, looking for evidences of Maya civilization. Travel adventures, Mexican and Indian culture, etc. Total of 669pp. 5⅜ x 8½.   20926-1, 20927-X Pa., Two-vol. set $7.90

AMERICAN LITERARY AUTOGRAPHS FROM WASHINGTON IRVING TO HENRY JAMES, Herbert Cahoon, et al. Letters, poems, manuscripts of Hawthorne, Thoreau, Twain, Alcott, Whitman, 67 other prominent American authors. Reproductions, full transcripts and commentary. Plus checklist of all American Literary Autographs in The Pierpont Morgan Library. Printed on exceptionally high-quality paper. 136 illustrations. 212pp. 9⅛ x 12¼.
23548-3 Pa. $12.50

CATALOGUE OF DOVER BOOKS

AN AUTOBIOGRAPHY, Margaret Sanger. Exciting personal account of hard-fought battle for woman's right to birth control, against prejudice, church, law. Foremost feminist document. 504pp. 5⅜ x 8½.
20470-7 Pa. $7.50

MY BONDAGE AND MY FREEDOM, Frederick Douglass. Born as a slave, Douglass became outspoken force in antislavery movement. The best of Douglass's autobiographies. Graphic description of slave life. Introduction by P. Foner. 464pp. 5⅜ x 8½. 22457-0 Pa. $6.50

LIVING MY LIFE, Emma Goldman. Candid, no holds barred account by foremost American anarchist: her own life, anarchist movement, famous contemporaries, ideas and their impact. Struggles and confrontations in America, plus deportation to U.S.S.R. Shocking inside account of persecution of anarchists under Lenin. 13 plates. Total of 944pp. 5⅜ x 8½.
22543-7, 22544-5 Pa., Two-vol. set $12.00

LETTERS AND NOTES ON THE MANNERS, CUSTOMS AND CONDITIONS OF THE NORTH AMERICAN INDIANS, George Catlin. Classic account of life among Plains Indians: ceremonies, hunt, warfare, etc. Dover edition reproduces for first time all original paintings. 312 plates. 572pp. of text. 6⅛ x 9¼. 22118-0, 22119-9 Pa.. Two-vol. set $12.00

THE MAYA AND THEIR NEIGHBORS, edited by Clarence L. Hay, others. Synoptic view of Maya civilization in broadest sense, together with Northern, Southern neighbors. Integrates much background, valuable detail not elsewhere. Prepared by greatest scholars: Kroeber, Morley, Thompson, Spinden, Vaillant, many others. Sometimes called Tozzer Memorial Volume. 60 illustrations, linguistic map. 634pp. 5⅜ x 8½.
23510-6 Pa. $10.00

HANDBOOK OF THE INDIANS OF CALIFORNIA, A. L. Kroeber. Foremost American anthropologist offers complete ethnographic study of each group. Monumental classic. 459 illustrations, maps. 995pp. 5⅜ x 8½.
23368-5 Pa. $13.00

SHAKTI AND SHAKTA, Arthur Avalon. First book to give clear, cohesive analysis of Shakta doctrine, Shakta ritual and Kundalini Shakti (yoga). Important work by one of world's foremost students of Shaktic and Tantric thought. 732pp. 5⅜ x 8½. (Available in U.S. only)
23645-5 Pa. $7.95

AN INTRODUCTION TO THE STUDY OF THE MAYA HIEROGLYPHS, Syvanus Griswold Morley. Classic study by one of the truly great figures in hieroglyph research. Still the best introduction for the student for reading Maya hieroglyphs. New introduction by J. Eric S. Thompson. 117 illustrations. 284pp. 5⅜ x 8½. 23108-9 Pa. $4.00

A STUDY OF MAYA ART, Herbert J. Spinden. Landmark classic interprets Maya symbolism, estimates styles, covers ceramics, architecture, murals, stone carvings as artforms. Still a basic book in area. New introduction by J. Eric Thompson. Over 750 illustrations. 341pp. 8⅜ x 11¼.
21235-1 Pa. $6.95

CATALOGUE OF DOVER BOOKS

GEOMETRY, RELATIVITY AND THE FOURTH DIMENSION, Rudolf Rucker. Exposition of fourth dimension, means of visualization, concepts of relativity as Flatland characters continue adventures. Popular, easily followed yet accurate, profound. 141 illustrations. 133pp. 5⅜ x 8½.
23400-2 Pa. $2.75

THE ORIGIN OF LIFE, A. I. Oparin. Modern classic in biochemistry, the first rigorous examination of possible evolution of life from nitrocarbon compounds. Non-technical, easily followed. Total of 295pp. 5⅜ x 8½.
60213-3 Pa. $5.95

PLANETS, STARS AND GALAXIES, A. E. Fanning. Comprehensive introductory survey: the sun, solar system, stars, galaxies, universe, cosmology; quasars, radio stars, etc. 24pp. of photographs. 189pp. 5⅜ x 8½. (Available in U.S. only)
21680-2 Pa. $3.75

THE THIRTEEN BOOKS OF EUCLID'S ELEMENTS, translated with introduction and commentary by Sir Thomas L. Heath. Definitive edition. Textual and linguistic notes, mathematical analysis, 2500 years of critical commentary. Do not confuse with abridged school editions. Total of 1414pp. 5⅜ x 8½.   60088-2, 60089-0, 60090-4 Pa., Three-vol. set $19.50

Prices subject to change without notice.

Available at your book dealer or write for free catalogue to Dept. GI, Dover Publications, Inc., 180 Varick St., N.Y., N.Y. 10014. Dover publishes more than 175 books each year on science, elementary and advanced mathematics, biology, music, art, literary history, social sciences and other areas.

3 1641 00120 7103

THOMAS CRANE PUBLIC LIBRARY

a31641001207103b

c.1

511.3 Stoliar, A. A.
STO   (Abram Aronovich)

Introduction to
elementary
mathematical logic

$5.00

APR 2 1987

© THE BAKER & TAYLOR CO.